THE MAGICK OF MATTER

THE MAGICK
OF MATTER

*Crystals, Chaos and the
Wizardry of Physics*

Felix Flicker

P
PROFILE BOOKS

First published in Great Britain in 2022 by
Profile Books Ltd
29 Cloth Fair
London EC1A 7JQ
www.profilebooks.com

A CIP catalogue record for this book is available from the British Library.

ISBN 978 1 78816 748 2
eISBN 978 1 78283 823 4

Text design by Crow Books
Printed and bound in Great Britain by Clays Ltd, Elcograf S.p.A.

Contents

The Physics of Dirt

The wizard, Veryan, whispered into her crystal the familiar incantation as she clambered through the cold, dark cavern. With a puff of breath, as if to release the seeds of a dandelion, she awakened in the stone a dazzling red light which illuminated the moss-covered rocks around her.

After walking for some time, she found herself at an entrance. Her passage was barred by a vast wooden door held together with wide iron joists. Working in the harsh light of the crystal, she felt her way to the door's handle, a thick, black iron ring. She pulled, but the ring held: the door was locked. Finding the edge of the wood, she pressed her fingers into the small gap between the door and its surrounding and found the door's bolt, made of the same rough iron as the handle.

She spoke again to the crystal in her practised, quiet tone and it gradually dimmed. After a few seconds, she found herself once again in perfect darkness. Positioning herself in front of the bolt, she rested the crystal flat on her palm, offering it to the door like grass to a horse. Uttering a few syllables of the old tongue, more sharply this time, the light returned as an intense red heat, a focused, narrow beam. Piercing the gap, the stream of light cleaved through the bolt, leaving in its wake the red-orange glow of molten iron, the noxious smell of the blacksmith's forge cutting through the air. The light vanished as quickly as it had appeared. She pulled again at the ring; with effort, the door eased open. As it did, light and life began to seep in through the widening crack from the dry stone staircase on the other side. Her task was about to begin.

This is a book about wizardry. It will reveal the secret ways of the wizard's art, and how you, too, can learn to follow them. It is also a history of magic, telling how, by a process of observing the world, wizards deduced the spells they cast – and how modern wizards continue to develop new magic to transform the world before our very eyes.

The modern name for magic is physics, and the name for a wizard's magic is *condensed matter physics*. Before we discuss what these names convey, you must understand that this book comes with a warning. Once you have learnt how a spell is cast, the effect of the spell will cease to appear to you as magic. It will become mundane. Everyday. *Boring.* This is the cost of magical knowledge. It will take a great deal of practice, and patience, for you to regain the sense of wonder you had when the magic was performed for you.

Throughout most of history – even within living memory – the story you just read would surely have been the stuff of fantasy. If you could produce from your pocket a crystal able to light a cavern at your request, then magic must be at work, and you must surely be a wizard. Yet these days such an action is mundane: an LED, a light-emitting diode, is a crystal, and by passing electricity through it you can cause it to light at the flick of a switch. A laser diode, also a crystal, creates an intense light which, when focused, can cut through solid metal. Now you probably feel cheated. There's no magic in using an LED torch. Using an LED torch is *boring!* Magic requires a certain incomprehensibility and unfamiliarity. LED torches are boring because they're familiar, and because, at one level or other, you understand how they work. But if you showed the torch to someone in the Middle Ages, they would certainly think it was magical because its technology would be unfamiliar. With enough time you could explain it to them. As you did, and as they gained familiarity, it would cease to appear magical. But would it really have lost its magic? Or is that just an illusion?

It takes work to see the magic in the familiar, but it's there. Physics is a programme to rationalise and understand the world. Many things that would once have been considered magic are now routine. Yet our understanding tends to advance in increments, building on existing knowledge. You may find a joke funny forever, but you can only 'get' it once. But understanding the joke allows you to perform it for others. With skill in the telling, and a little luck, a joke will have on others the effect it once had on you. So it is with magic. The secret to learning the world's magic – to learning physics – is to laugh continuously at the cosmic joke. It's the difference between seeing a conjurer perform a trick and having the trick explained. Hopefully when you first meet some of the ideas in this book, they may invoke in you that sense of magic. And hopefully, when you've read the book, you will understand where those ideas came from, and they will seem more natural. You may have to work to maintain the sense of magic they once held, but by learning a spell you can cast it to the benefit of others.

The rules of wizardry

Now that you're heeding the warning, let's talk about wizards. When I refer to wizards, I'm thinking, like, classic wizards. People who do magic.* I'd say the defining characteristics of a wizard are something like this. We can call them the Rules of Wizardry:

1. A wizard studies the world.
2. A wizard understands that they are a part of the world they are studying.
3. A wizard's understanding leads them to see hidden patterns and connections that others do not.
4. A wizard's knowledge is of a practical, hands-on kind.
5. A wizard can cause changes to the world, but they make such changes sympathetically (see Rule 2).

Sometimes a wizard's study is academic, like Harry's and Hermione's at Hogwarts. Sometimes the study is a quiet contemplation, as with Rey or Yoda in *Star Wars*, sages in classic Daoist texts such as the *Zhuangzi*, or martial artists such as Katara and Aang in the epic TV series *Avatar: The Legend of Aang*. Often the study takes the form of exploring and experiencing the world, as with Gandalf in *The Lord of the Rings*, Morgana and Merlin in Arthurian legend, and Tenar and Ged in Ursula K. LeGuin's classic *Earthsea* novels. In many modern examples, the wizard is a supernaturally gifted scientist. Doc Brown's achievements in *Back to the Future*, Rick's in *Rick and Morty* or those of Doctor Who are presented as scientific, but the technology is so far beyond the experiences of the other characters and the audience that it is more like magic. It is apparent that the Rules of Wizardry implicitly assume an important hidden rule, *The Rule of Rebellion*:

A wizard understands that rules are made to be broken.

---

* I have tended not to use the term 'witch' because of its historical associations with persecution. The term 'wizard' is intended to cover magic workers of all types and backgrounds.

Look again at the five-and-a-half rules of wizardry, and replace the word 'wizard' with 'scientist'. That seems about right, doesn't it? J. G. Frazer, in his chronicle of magical practices *The Golden Bough*, put it poetically:

> *Magic like science postulates the order and uniformity of nature; hence the attraction both of magic and of science, which open up a boundless vista to those who can penetrate to the secret springs of nature.*

Frazer's quote makes a link between magic and science. But wizardry is a particular type of magic; a particular type of science. And its name is condensed matter physics.

The magic of true names

Zoology is the study of animals. Botany is the study of plants. What is physics the study of? Its name derives from the ancient Greek, *ta physika*, meaning 'natural things', taken from the title of Aristotle's collected works on the physical world. That doesn't narrow things down much. Perhaps the best answer is that physics is defined not so much by the set of phenomena studied, but rather by a distinctive approach and set of tools. Broadly, these tools divide into three groups. A physicist will tend to specialise in just one of them, although it takes all three working together to obtain the desired knowledge of natural things.

The three categories of tools are experiment, numerics and theory. Experimental physicists – experimentalists – carry out practical tests to see how the world behaves. However exotic and unfamiliar our scientific theories may become, they must always lead to testable predictions. These predictions can be confirmed or falsified through observation: wizards don't invent spells, they learn them.

When our fictional accounts of wizards give us glimpses of how wizards learn their spells, it is invariably through observation of the world itself. In *Avatar*, for example, certain individuals are born with a magical influence over water, which their ancestors learnt from observing the moon's influence on the tides.

Numerical physicists – numericists – build and test computer simulations of the world. Simulations can be conducted under more controlled conditions, and repeated more frequently, than if the experiments were carried out in reality. The trade-off is that numericists need to know that their simulation shares all the relevant properties with its real-world counterpart.

Theoretical physicists – theorists or theoreticians – also work with models of reality. But whereas numericists would generally prefer the most accurate simulation, theorists generally seek the simplest model that captures the essence of the phenomenon. A theorist must learn to see through to the true essence of a thing; this process surely lies at the heart of all magic.

I am a theorist, although I also work closely with experimentalists and numericists. This guide to modern wizardry will present a theoretical physicist's perspective – partly because this is the perspective I have, and partly because it is in the nature of a book such as this to boil complex stories down to their essence. Theorists build analogies; fables. But as a fellow theorist Dr Jans Henke once put it to me, mathematical models are the most powerful kind of analogy, because they don't just relate phenomena to familiar cases: they also allow us to say, in detail, how they will behave in new, untested, situations. Experimentalists can then test the phenomena and see if they behave as the model predicted. Often the experimental observation comes first and the fable is woven around it. Suppose a model's prediction is verified experimentally in a repeatable, controllable way. That lends weight to the idea that the simple elements which went into the model captured the essence of the phenomenon. Theoretical physics often comes close to mathematics; the difference enters via the gap between the mathematical model – perfect and predictable – and reality, the messy world we experience. Theoretical physics is the storytelling we do to make the mathematical model more intuitive.

The work of the theorist always reminded me of the magic of true names. From ancient Egypt to modern hacking culture, the idea has persisted that learning something's true name grants us power over it. Ursula K. LeGuin's *Earthsea* books, which are said to be the first example in fiction of the wizard being the protagonist rather than a supporting character, provide a great example in a fantasy setting. In the world of Earthsea, wizards gain their magic by listening to the world and learning the true names of things. Now, the day-to-day names we use for things are simply labels we attach so we can refer to them in conversation. In Earthsea these are called use names, but things also have a true name. These names are said to belong to the *Language of the Making*. We are told the true name for pebble is 'tolk', for example. When we say something's use name to someone else, a little bit is lost in translation. When I say 'pebble', I conjure certain associations in my mind that other people will not have. My fiancée Dominique explained it to me like this. If you were to speak something's true name, then by definition nothing could be lost in translation; anyone would have the same perfect understanding of what is meant. So it is natural to associate true names with conjuration. How can you guarantee perfect understanding unless the thing itself is present? When I say 'pebble', I may be referring to some more general property shared by all pebbles. To speak the name 'tolk', however, I must first have understood the essence of a pebble.

People, too, can have true names. In the graphic novel *The Invisibles*, a person must take a magical name when they become a wizard. Grave warnings are issued against flippant name choices because the name shapes their personality. I have a friend who belongs to a religion in which a holy person has to be consulted when naming babies. They are believed to have some mystical understanding of the essence of the child, and name them accordingly. This holy person then assigns new names as the person grows throughout their life. It is true that names can dictate elements of one's life. My own name, Felix Flicker, is absurd and demands attention; I can't help but wonder whether I internalised those traits. The effect can be more serious, though. A 2012 study found that when identical applications were assessed for a scientific job, the application was deemed of lower quality when a female name was attached to it than when a male name was attached, and a significantly lower salary was deemed appropriate.[1] Even in our world, a name is more than an arbitrary label.

Theorists study models of things, not the things themselves. Say that one day a theorist drops her crystal ball down the steps of her tower. Imbued as it is with potent magic, the ball will survive intact. But she needs to know when it will arrive at the bottom, in order to summon an eagle to collect it in a timely manner.* Quick as a flash, she decides to use Newton's laws of motion to construct a mathematical equation to model the ball's descent. But she will not attempt to capture every feature of the physical scenario. Probably she will assume the stairs are frictionless; probably she will ignore air resistance; probably she will ignore little gusts of wind which might come about, because these can't be predicted with any certainty. Our theorist hopes that the outcome – which she can calculate with certainty in her model – matches the reality in which she has found herself by tossing her orb about. The crystal ball is the use name – this particular ball – while the mathematical model is the true name: perfect, and untainted by reality. Once you understand

---

* Eagles, as Tolkien taught us, are always happy to oblige with a convenient rescue.

a piece of mathematics, you understand it in exactly the same way as anyone else who understands it, regardless of what language you speak. Two plus two equals four however you write it. There is no approximation in a model; the approximation appears in getting from the model to reality. It is a source of much philosophical debate as to whether the model 'exists'. If it does, then it might not be too much of a stretch to suppose that understanding the model conjures it into that existence. To view the world as a theoretical physicist, you must learn to listen for the true names of things: you must learn to conjure perfect mathematical models. The art lies in choosing the simplest models that capture the essence of the thing being studied. This simplicity is important; a map with a 1:1 scale would be entirely accurate, but it would also be entirely useless because it would give no simplification.

Physics is a set of tools that can be applied to anything, from the invisibly small to the unknowably large. But the wizard's focus is more specific, lying in the here and now. Between the extremes lies a middle realm: the familiar world we inhabit.

The middle realm has its own ways

All disciplines of physics do pretty magical things. Cosmologists study the birth and life of the universe, and also predict its fate. Astrophysicists have listened to gravitational waves to hear black holes collide. Particle physicists excite quantum fields to create elementary particles that have never before been detected. These are very grand magics, and many excellent books have been devoted to them. Yet between the microcosm of the quantum and the macrocosm of the universe lies a middle realm. It is no less magical, but its magic takes a different form – a familiar form – and as such it has been largely overlooked in popular books. Yet it is the largest area within physics, occupying around a third of all researchers.

The study of the middle realm is condensed matter physics. It is the physics of the things you see around you: matter – lumps of stuff you can hold in your hand – and their description, right down to the quantum realm from which they emerge. Wolfgang Pauli, one of the creators of quantum mechanics, famously dismissed condensed matter physics as *Schmutzphysik* (the physics of dirt). It's the perfect description of the wizard's art.

I think it is fair to say that condensed matter physics' closest cousin is particle physics. It is important to understand the similarities and differences between these disciplines. Particle physics is the study of elementary particles – electrons, protons, and so forth; a reasonable definition might be something like this:

An elementary particle can exist by itself in the vacuum of space, and cannot be reduced to other things with that property.

An electron meets these criteria. An atom, however, does not, because while it may be able to exist by itself, it is made up of other things (electrons, protons and neutrons), which also have that property. Protons are themselves made up of three quarks; but quarks cannot exist in isolation, so by the definition above they are also not elementary.

Now, condensed matter physics is the study of what emerges when many elementary particles interact. If that's so, doesn't it simply reduce to particle physics? In this book I'll try to convince you that the answer is no. If condensed matter physics had a tagline, it would be this:

The whole is more than the sum of the parts.

Perhaps the most important illustration is given by the behaviour of particles within matter – to my mind the central set piece in the magic show of reality. When an electron shoots through the vacuum of space it has a specific mass, charge and magnetic field (called its 'spin'). These uniquely define it to be an electron, and all electrons are alike. If that electron travels into a material, it interacts with the other particles in the material according to the rules of quantum mechanics. In doing so, its properties change; since all electrons have the same mass, it can no longer be an electron. Indeed, it is no longer an elementary particle: it has transformed into an 'emergent quasiparticle', the whole which is more than the sum of its parts.

To explain how this works, I will rephrase an elegant analogy devised by Professor David J. Miller to explain the behaviour of the Higgs boson, an elementary particle. Miller mentioned that he had borrowed the central conceit from condensed matter physics, so I trust he won't mind a temporary return loan. Imagine a collection of avid ghost hunters has packed into the dilapidated ballroom of a haunted mansion, unbeknownst to the ruff-wearing spectre who is happily floating down the corridor with his detached head held under his arm. The ghost enters the ballroom, and suddenly all eyes (and dubious measurement devices) are on him. The crowd, previously spread out, squashes around him. Unfortunately for the ghost, he's the kind from *Tom's Midnight Garden* which can't pass effortlessly through people. His pace dramatically slows as he has to push his way through the crowd of ghosthunters failing to capture him on camera. The ghost's mass has increased, in the sense that it would take a greater force to accelerate him than when he was strolling alone down the corridor: he now has a surrounding crowd which also needs to be moved. To bring the analogy a bit closer to reality's true quantum weirdness, we might imagine that he is instead the kind of ghost from *Bill and Ted's Bogus Journey*: rather than push through the crowd in his original form, he hops between host bodies as he possesses them one after the other. He again slows down and effectively gains mass, but there is now nothing in the ballroom resembling the original ghost at all; yet when he pops out onto the veranda he reappears in his original form. When the electron is in the material it is changed, yet it can leave the material and return to being an elementary particle.

Other emergent quasiparticles have no precedent in the world of elementary particles. For example, while light is conveyed by elementary particles called photons, sound cannot be described by elementary particles, for it cannot exist in the vacuum of space. Sound, being a vibration, requires a medium through which to travel. Yet it *can* travel through matter – and when it does, it too can be described by emergent quasiparticles, known as 'phonons'.* To borrow again from Miller's analogy, this time a ghosthunter merely imagines they 'have felt a presence' and tells the person next to them. That person's neighbour overhears and leans in, and soon the rumour is moving around the room. Wherever the rumour goes, the crowd squashes together *as if* there were a ghost there – but there's not. This dense region of crowd behaves like an object with mass, resisting changes to its motion, just as a phonon does. Matter that doesn't contain any quasiparticles can be thought of as the condensed matter version of the vacuum of space – after all, a vacuum is simply the absence of elementary particles. Phonons provide an illustrative example. They can be understood as the vibrations of the atoms in a crystal; when the crystal is cooled down, the atoms vibrate less and the phonons disappear. When all phonons are gone the crystal is in its lowest energy state, called its 'ground state'. Were you to speak to the crystal in your quiet, practised tone, you'd give it energy, causing its atoms to vibrate and conjuring phonons into existence. This motivates the following definition:

---

* Some physicists prefer to call phonons 'collective excitations' rather than 'quasiparticles'. I will not draw this distinction.

An emergent quasiparticle can exist by itself above the ground state of a material, and cannot be reduced to other things with that property.

Emergent quasiparticles cannot be reduced to elementary particles without losing an essential part of the description: think of the crowd squashing together to hear the rumour of a ghost. It's true that everything can be described in terms of individual ghosthunters, but that would miss the bigger picture. This is the essential idea of 'emergence', the concept that the whole can be more than the sum of its parts: the crowd has properties, such as moving constrictions, that are not properties of the individuals who comprise it. In condensed matter physics the individuals will usually be the atoms and elementary particles, and the emergent properties will be the large-scale behaviour of matter, understood in terms of emergent quasiparticles.

Quasiparticles are unique to condensed matter physics. Many have a dreamy sense of unreal wonder: phonons can be measured in experiments – but if you look for them at the level of elementary particles, there is nothing there. They are simply the collective vibrations of atoms.

Now, it may seem tempting on that basis to dismiss quasiparticles as less 'real' than the elementary particles from which they emerge. But look at it like this: we consider the world around us – the middle realm – to be real. By contrast, we think the quantum realm from which it emerges is full of mystical hocus-pocus. Yet our familiar world only avoids quantum tomfoolery *because* it is emergent. To reject the reality of quasiparticles is to reject the reality of everyday existence. There is no elementary particle which carries sound, yet you are able to hear the distant hoot of owls in a nearby wood.

The owls are not what they seem

Theoretical physicists can be found in all branches of the subject, boiling reality down to its essence. But this boiling down can take many forms. Particle physicists are trying to identify the individual building blocks of the universe – the smallest moving parts of reality. This programme has had phenomenal success, culminating in the Standard Model which accounts for all known elementary particles. Perhaps the ultimate aim of this quest is the 'theory of everything', which would add the final missing ingredient to the Standard Model: gravity. If found, this theory would then encompass all the forces in nature. It would explain dark matter and dark energy, and it would contain within it the key to understanding the fate of the universe. But you can probably see that it wouldn't really be a theory of *everything*. In fact, it would not really describe *anything* you actually experience day to day. It would be a theory of all the elementary particles and their interactions, but it would not be a theory of, say, owls.

There is no elementary owl particle, yet we believe owls exist. They are made up of many different types of atom. Each atom is made up of protons, neutrons and electrons. So owls are not elementary; they are emergent. Owls are complex, messy, sets of traits; they are more than the sum of the parts from which they emerge. The simpler parts could be elementary particles, atoms, cells, genes or other things. These lower-level descriptions are not mutually exclusive, and none of them is wrong, as such. But they do not account for the owl's talons, its screech, its beak or its popular association with magic.

Condensed matter physics is not the study of owls (at least, not yet). But it is the study of what emerges when many things interact, and this is what distinguishes the middle realm from the microscopic world. A well-worn saying has it that two heads are better than one. What is less frequently observed is that two heads are more than twice as good as one: the extra bit is emergence. And when many, many particles get together to form a lump of stuff, new worlds can emerge.

This book asks what form these new worlds can take. If you were to boil the story down to its essence, the tincture you would create would provide an answer to the following question: what is matter?

There are many ways to understand the answer, and we will encounter a complementary approach in each chapter. Ours will be a journey of discovery, undertaken in three stages.

There and back again

Once, when walking through the desert, I chanced upon a magician. Our conversation drifted this way and that, and we found our way to the subject of stage magic. I asked the magician if he knew of my favourite performer, Derren Brown, the master illusionist who reminds us at the start of each show that he employs a combination of 'magic, suggestion, psychology, misdirection and showmanship'. The magician knew of him.

I told the magician that my appreciation of Derren Brown had grown in two stages. When I first watched him, I enjoyed the brilliant feats of mentalism. He can read your mind, he can make you see things which aren't there, he can make you perform impossible stunts. He exploits gaps in our perceptions of reality to show us that our models of the world are susceptible to manipulation. But after obsessively rewatching his act, I came to realise that a lot could be accomplished using clever sleight of hand and traditional magic, rather than psychological manipulation. This was when I reached the second stage of my appreciation. This was the *real* magic trick! He made a rational sceptic like me believe in magic again, and he accomplished this by appealing to a scientific blind spot: the mysteries of the mind.

The magician in the desert told me that he considered Derren Brown to be the greatest living magician. He agreed that, as with all truly great magic, there exist the two stages of appreciation I'd glimpsed. But he added that there was also a third stage: that of the professional colleague watching the tricks. A magician will know many of the techniques being used and how the tricks are performed – but it is still a marvel to watch Derren Brown working because his technical ability is unparalleled. It is a joy for a professional to see the tricks performed so deftly.

I heard in the magician's words a parallel to the scientist's journey of understanding. It is a three-stage journey that we have all begun. When we are young, we are fascinated by the world. It is all new to us, and we marvel at its wonders. This is the first stage: enjoying the performance. As we get older, we begin to learn how things work. We approach the second stage: understanding how the world performs its magic. It is easy to get lost here, stumbling into a cold, dark cavern of rationality. But if you can keep that fire of excitement smouldering inside you, it takes little more than a short puff of breath to reignite it. With patience, and a little luck, you can kindle that fire and proceed to the third stage of understanding: that of the scientist, who understands how the world works its magic and loves it all the more for the skill of its performance.

Our understanding of the stuff around us is a story that has been updated and retold through the ages. As we progress through this book, we will meet different takes on the nature of matter. We will begin in the distant past, where all was earth and air and fire and water, and we will progress toward the far future, in which our lives are transformed by things condensed matter physicists are only now beginning to comprehend. In the earlier chapters, we will put together our essential spells: the knowledge, passed down through generations, representing the understanding that all condensed matter physicists must develop in their training. Thus prepared, in the later chapters we will push on to the future, meeting spells condensed matter physicists are still learning to cast.

If there is one message you take away from this book, I hope it is this. Wizards are real, and if you are interested in becoming one, the condensed matter physics community will welcome you. If you are concerned that you do not fit a traditional idea of a physicist, you are needed all the more. There are condensed matter physicists from all walks of life, and increasingly so. I will give a snapshot of contributions to the subject made by some individual physicists past and present, and I hope these give some idea of the breadth of backgrounds of people behind the science.

On the other hand, any small insight into the characters of the past should not be taken as an endorsement of all they have said. J. G. Frazer, quoted above, had an elegant turn of phrase, but his views were regressive even for their time. And there are battles still to be won. For example, until 2017, only two of the 216 physics Nobel laureates were women; it took thirty years for the first prize to be awarded outside Europe or North America, and to this day it has never been awarded to anyone from Africa, South America or the Middle East. Nobel Prizes are merely one symptom of a much wider problem in which only a narrow cross-section of society is encouraged to pursue science, and in which the contributions of people who don't fit the stereotype are devalued or ignored. I will occasionally highlight Nobel Prizes as a convenient shorthand to indicate the importance of certain work, but it should be understood that the lack of a prize is often an indicator of nothing more than bias. Things are improving: the physics prize was awarded to women in 2018 and again in 2020 (although seven men won it in the same period). This improvement is vital for the future of science: the best way to solve a complicated problem is to have as many views and approaches represented as possible. So if you've ever felt excluded by past depictions of scientists or wizards, it is because you are the future of the subject.

I wish I could tell you that I was inspired to study physics by a desire to be a wizard, or a love of fantasy fiction. I do recall being drawn to the arcane words and the idea of an esoteric knowledge available to the initiated, but it was really something deeper that drew me to both: a love of imagination. The same power which is used to create imaginary worlds is used to realise those worlds in physical theories, and to invent ways to access them in experiments. I will employ this connection throughout the book, using both fictional passages, and references to classic books and films, to emphasise the magic behind the physics. As with the opening passage of the book, it is often easier to see magic when it is presented as fiction; but by the close of the book I hope you will agree that the real world is as magical as the most enchanting tales it contains.

Let us proceed, then, with learning the physics of dirt. There are many spells that don't appear in this book, and it is not the purpose of this book to teach you them. The world is already telling you its spells; the purpose of this book is to help you to listen.

II

The Four Elements

One day, Lady Long-Ears and Mister Calabash were beating olives from a tree. A swarm of mayflies surrounded them. Mister Calabash said: 'Have you ever considered that these mayflies spend their entire lives with us, Long-Ears? We arrive at our work at dawn. By our morning tea break they are born, by our afternoon lunch break they are middle-aged, and by the time we walk home down the mountain they are dead. The lucky ones mated, but many more pass their brief and meaningless lives unnoticed, proving at most an occasional annoyance to us at our work.'

Long-Ears replied: 'Their lives are no more meaningless than yours or mine. Consider this olive tree, for example. It sprouted in the time of our long-forgotten ancestors. By the time it had matured, many generations of our families had lived and died. When its time is up, you and I will have passed into obscurity.'

To this the olive tree added: 'Further to your point, you two will have proven little but an occasional annoyance to me, too.

'The first people to find me left me alone. Their children's children ate my fruit and found them disgusting. Their grandchildren learnt to salt those fruit. Since then I've been picked at incessantly. It was only another hundred years or so before you started growing my offspring, down there. Another generation and you were watering them properly, although not as properly as this mountain waters me.'

The mountain joined in: 'I should think not.

'Your observations, olive tree, are insufficient. To me you are all insignificant. Tree, I have watched animals evolve over countless generations to eat your fruit, just as I watched your species evolve to use the animals.

'In my youth I came from the sea and watched continents collide. As I matured I watched species come and go. To me, your entire species are mere fleeting specks. I live on the timescale of our entire planet, and understand that of the universe itself.'

The universe felt compelled to join the discussion at this point.

'Mountain, you speak of understanding my timescale, but you do not grasp that I have no such thing. I stretch long into your future and long into your past, but I also inhabit every event in between. You see the species on you come and go, but the generations are of so little consequence to you that you cannot discern them.

'Tree, I enjoy your set of events immensely. You too are proud in your way. You see the generations of the likes of Long-Ears and Calabash, but they themselves provide you with only a fleeting glimpse of their existence.

'Mister Calabash, you claim the mayflies' entire lives are insignificant compared to a single uneventful day of your own. Perhaps you would consider the following.

'You are all correct and all incorrect, as am I. My true nature cannot be conveyed, but I will give a crude approximation. I join each of you on your timescales, but I also contain every other event that has been and which ever will be. In your understanding, these things are happening together.

'I contain the annihilation of particles so short-lived they cannot truly be said to have existed at all, and others which exist forever without ever meeting another thing. I contain the mayflies living out their days and the planets living out theirs, and all these things are both absolutely significant and completely pointless.'

With that, Mister Calabash took back his comments about mayflies and went home and had a good hard think about what he was going to eat that evening.

❧

Emergence

How many grains of sand does it take to make a pile?

There's no obvious, satisfying answer. You could suggest a number: four, say. But four isn't a very convincing pile, and if anyone asked why three isn't a pile and five is, you can only answer that you've given your number and you're sticking to it. This is what scientists did when defining the number of atoms in a reasonably sized lump of stuff. The number, called Avogadro's constant, is precisely

602,214,076,000,000,000,000,000.

It's just over half a trillion trillion. That number is huge: by comparison there are a mere two hundred billion stars in the Milky Way. But huge as it is, Avogadro's constant is nevertheless a fixed number: one atom fewer and you're a little shy of a reasonably sized lump of stuff, according to the definition.

Maybe a vague answer is better than a precise one. It's often easy to say whether or not a given set of grains constitutes a pile, even if a precise definition is difficult. The ambiguity as to where the shore lies does not change the fact that it's easy to point to something that is definitely the sea and something that is definitely land. Two grains of sand is definitely not a pile, but a million grains of sand all in contact and arranged into a vaguely cone-like shape definitely is: the pile has emerged from the grains.

This chapter is all about emergence: about how interactions between atoms on the smallest scales of length and time lead to large-scale measurable effects in our middle realm.

Ours is a realm built from states of matter – that is, from atoms behaving collectively. To understand emergence is to understand how the states of matter grow out of the world of atoms. I walk on the earth, swim through the water, breathe the air, and when I am done I warm myself by the fire. This may seem no stranger than the pile of sand emerging from the grains. But the very same atoms can exhibit many different emergent behaviours; for instance, water molecules can combine to form ice, water or steam. There are many such subtleties to grapple with. To understand the states of matter and their transformations, we will have to think across different scales of length and time, as Mr Calabash learnt to do; and, like Calabash, we will come face to face with structures that appear on all scales of length and time, from small to large, simultaneously. Our quest will take us to the distant past, granting primordial glimpses of condensed matter physics, via early ideas of emergence. It will be a most wizardly initiation.

Now, there is no denying that wizards are hoarders of arcane knowledge. But they do not hoard for hoarding's sake: they always have a practical use in mind (see Rule 4: *A wizard's knowledge is of a practical, hands-on kind*). By better understanding the states of matter, we can better turn them to our advantage. This does not mean bending them to our will: we are still a part of the world (Rule 2: *A wizard understands that they are a part of the world they are studying*). The idea is put nicely by the *Book of Liezi*, a Daoist text from fourth-century CE China. It tells of Confucius witnessing a swimmer passing effortlessly through a seemingly impassable whirl-pool. The swimmer explains:

> I enter the vortex with the inflow and leave with the outflow, follow
> the way of the water without imposing a course of my own.
> Book of Liezi, translation by A. C. Graham

The swimmer understands that they are part of the water's flow, and benefits by going with it. While disciplines such as particle physics, cosmology and astrophysics might reasonably draw a distinction between the scientist and the subject of their study, condensed matter physics is inherently linked to the world of everyday experience. This practical side of arcane knowledge has been a part of condensed matter physics from its earliest prehistory; and it was in the forefront of the minds of the two people who granted the subject its name.

More is different

Condensed matter physics was summed up by one of the field's progenitors, Philip Warren Anderson, in a simple phrase: *more is different.* This is the essence of emergence. In a 1972 article of the same name, he argued that:

> The ability to reduce everything to simple fundamental laws does
> not imply the ability to start from those laws and reconstruct the
> universe. In fact, the more the elementary particle physicists tell
> us about the nature of the fundamental laws, the less relevance
> they seem to have to the very real problems of the rest of science,
> much less to those of society.

Within the Cavendish Laboratory in the University of Cambridge, where Anderson spent much of his career, there was once a research group by the name of 'Solid State Theory'. It had been apparent to the members of this group for some time that their interests ranged well beyond solids. In 1967 Anderson and his colleague Professor Volker Heine renamed the group 'Theory of Condensed Matter', thereby acknowledging all matter that has in some sense 'condensed', meaning that the interactions between the particles have resulted in a collective behaviour on the everyday scale. Naming is an act of creation: it asserts that some things are to be included while others are not. Sociologist Mary Douglas, in her study of ritual taboos and magical beliefs *Purity and Danger*, put it succinctly:

> *As learning proceeds objects are named. Their names then affect the way they are perceived next time: once labelled they are more speedily slotted into the pigeon-holes in future.*

The name *condensed matter physics* expanded the subject's scope from solids to all of matter, an act that has shaped the development of the field to this day.

I had the good fortune to meet both Heine and Anderson. I met Heine in 2015, when I was finishing my PhD. I visited Cambridge to give a seminar, and on a whim I knocked on his door unannounced. He appeared before me wearing a psychedelic shirt and a large medallion necklace, enthusiastically welcoming me in despite having no idea who I was. His office was adorned with a lifetime's collection of otherworldly artefacts, and our discussion was one of the most interesting of my life. It started with physics, but we found our way via myriad subjects to his childhood escape from Nazi Germany to New Zealand. Similar escapes prove an unfortunately common theme in the history of twentieth-century physics.

The following year I began a research fellowship at the University of California, Berkeley. During this time I flew to Princeton to give another seminar, where I met Anderson, who had been based in Princeton since his retirement in 1984 (like wizards, physicists never really retire). The visitors' office was directly opposite Anderson's and his door seemed always to be open, so I knocked and said hello. Doing my best to maintain my composure in the presence of both Anderson and his life-size cut-out stood directly next to him, I was honoured with another fascinating conversation. It was deeply reassuring to hear Anderson's account of how he had reached his now-hallowed ideas. I think my comfort derived from realising that legends are people too; they had their own uncertainties about whether their ideas would be well received.

Later that day I mentioned my discussion with Anderson to some of the postdoctoral researchers in Princeton. They were shocked – they had never heard of anyone speaking to him! Because no one approached him, he was believed to be unapproachable. On the contrary, his enthusiastic encouragement proved a great help in what can often be an uncertain career.

The shoreline of condensed matter physics may be ambiguous, but all wizards can point to the sea of emergence as one of the subject's defining features. Nevertheless, despite being an intuitive idea, it can be hard to pin down in words. As a starting point we can borrow a famous Zen Buddhist *koan*:

> *This [claps] is the sound of two hands clapping. What is the sound of one hand clapping?*

The clap is not present in either hand; it emerges when the two combine. My philosopher friend Dr Leonid Tarasov suggested for us the following definition:

A phenomenon is emergent if and only if it is explainable in terms of other phenomena but can't be eliminated from the explanation in favour of them.

That is, to eliminate the emergent phenomenon would be to miss something essential. It is worth noting that emergence in this sense does not contradict that other mainstay of scientific thought, reductionism, which seeks to boil ideas down to their essence. Reductionism is the process of working out which bits of a story are actually important and which are irrelevant. When Sherlock Holmes solves a case, he repeats back to us the story we've just read, but in a reduced form which contains only the pertinent details. He is able to do this because he has understood the problem and knew which details to disregard. As my friend and fellow theorist Dr Chris Hooley likes to point out, emergence is actually a *form* of reductionism – it's just that the relevant details are not the smallest things (elementary particles) but collective phenomena. And this is quite intuitive: if you were given a few seconds to sketch a wizard, you would probably jot down a stick person with details such as a wonky hat with stars on it, a staff, maybe an owl on the shoulder. You would probably not set to drawing as many atoms as possible in your allotted time, even though the wizard can be described in such terms.

An intuitive explanation can be found in terms of ants. It's probably fair to say individual ants don't come up with particularly elaborate schemes, yet a colony of ants is collectively able to deduce quite clever things. The physicist Richard Feynman discusses his observations of ants at some length in his autobiography *Surely You're Joking, Mr Feynman!* First he notices that if you watch a trail of ants going to and from a source of food, such as a sugar cube, they're often taking a very efficient route. But how does an ant know the best route? On the ant's scale, the cube is far from the nest, and the ant presumably can't see or smell the cube from far away. Feynman observed the following. An ant will find the sugar cube fairly randomly. When it does, it collects some and finds its way back to the nest by a bit of a roundabout route. Feynman suggested that maybe the ant leaves a scent on its return journey which lets other ants know that it's onto a good thing, because then other ants start retracing the route to the sugar cube. The routes followed by later ants become increasingly efficient, as they cut corners and take short cuts. In no time at all there's a trail of ants describing a good approximation to the shortest route from the nest to the sugar.

Feynman had observed this natural phenomenon, and had come up with a theory as to why things were that way. But because he was a good scientist, he came up with a test for his hypothesis and checked it against reality. The ants were coming in through a gap near a window, arriving on a windowsill. He suspended the cube from a string so that an ant would be very unlikely to accidentally stumble across it. Then he placed a piece of paper on the windowsill. Whenever an ant got onto the paper, he transported the paper to the cube. Whenever the ant left the cube and got back on the paper ferry, it would be returned to the windowsill. In no time at all, the ants were forming a direct route to the paper ferry, riding it over to sugar cube, back to the ferry, and back to the nest. It confirmed the hypothesis about how the ants were working out the routes to take.

No individual ant came up with the understanding of how to use the ferry: the idea emerged from the set of ants collectively. In the wild, ants have been known to cling together to bridge gaps ten or twenty ant-bodies wide. On the other hand, their behaviour sometimes goes wrong: army ants have sometimes been found marching in 'death circles', in which huge numbers of them end up following one another in a circle until they eventually die of exhaustion. Establishing how this complex behaviour comes about from simple rules promises applications from 'swarm robotics' (simple robots working collectively without a leader) and nanotechnology to 'programmable matter' (whose molecules can be 'taught' to adjust their positions to useful effect). A major focus of computer science is artificial neural networks, in which a computer recognises patterns by using many simple processes in a collective way, inspired by neurons in the brain. In each of these cases there is a complex behaviour on larger scales which, while explainable in terms of simpler parts on a smaller scale, can't be eliminated from the explanation.

Perhaps the classic example, though, is the one we are most interested in here: matter itself.

States of matter

The states of matter were nicely characterised by the Greek philosopher Empedocles, who proposed that everything is made from some combination of the classical elements: earth, air, fire and water. Remarkably similar theories of matter existed across many cultures, including ancient India, Egypt, Babylonia and Tibet, as well as in Hinduism and Buddhism.

The idea may in fact have its origins in western Persia, and the priests of Zoroastrianism. The Magi, as they were known, lent their name to the word for magic in many languages due to their esoteric studies in alchemy, astrology and astronomy. It would seem reasonable to imagine them as ancient precursors to modern scientists; if so, the four elements grant primordial glimpses of condensed matter physics.

These elements have carried through remarkably well to modern science as the four familiar states of matter: earth is a solid; water is a liquid; air is a gas; and fire is a plasma, an example of the fourth state of matter. These states all have different properties, but the one thing they have in common is that they only emerge once you have enough particles that the individuals disappear into the crowd. To understand this, it is important to think about the world on different length scales.

Considering different scales of length and time is vital to condensed matter physics, because the subject routinely builds up from a description in terms of elementary particles all the way to the emergent properties of the everyday world. These scales can conveniently be grouped by the experimental techniques used to probe them.

For instance, take a look at your wizard's staff leaning against the wall. The object you see lives on the 'macroscopic scale', or macroscale, the familiar scale of everyday objects: things which can be seen with the naked eye. Lengths of, say, a few metres or more down to a millimetre.

If you have a microscope to hand, you can inspect your staff more minutely, down to lengths of around a thousandth of a millimetre. This is the larger end of the 'microscale'. Under a microscope you can see the individual plant cells of your staff; in the unlikely event that you have a scanning tunnelling microscope to hand (they tend to take up most of a room) you can see all the way down to the nanometre scale (a thousandth of a thousandth of a millimetre), or 'nanoscale'. Only about five of your staff's atoms would fit along a nanometre; a DNA helix is about three nanometres wide. The best scanning tunnelling microscopes can resolve about a tenth of a nanometre, roughly the diameter of a single atom.

When I heard it was possible to image individual atoms, I did not believe it. Surely the universe must conspire to keep such knowledge hidden. Yet nowadays I have the privilege to work alongside experimentalists who effortlessly peruse the nanoscale to learn the secrets of our world. One of the world's leading scanning tunnelling microscopists is Professor Vidya Madhavan at the University of Illinois in Urbana-Champaign. She and her PhD student Jorge Olivarez Rodriguez kindly sent a picture (Figure 1 on page 30) of strontium atoms they took using a scanning tunnelling microscope. The reason it is slightly blurry is that the microscope is hitting against fundamental limits imposed by quantum mechanics.

1: Individual strontium atoms in a crystal as seen using a scanning tunnelling microscope. Courtesy of Vidya Madhavan.

The phrase 'microscale' is sometimes used for all length scales requiring any kind of microscope to explore. That is the convention I will adopt here, meaning the world divides into two: the microscale and the macroscale. Equipped with this terminology, we can grasp the four elements and their respective states of matter.

Solids are represented in the four elements by earth. It's generally pretty intuitive what a solid is, but coming up with a precise definition turns out to be rather tricky. The definition that scientists have settled on is probably not the first that would come to mind; only solids, they determine, can resist 'shear stress'. A shear stress is the force created by pushing one surface in one direction and the opposite surface in the opposite direction, like a magician sliding playing cards off a deck. Imagine a rival conjurer snuck in and swapped the deck for a replica made from a single solid piece. The magician would not be able to slide the cards off because the atoms in the replica cards would be bonded together.

Solids divide naturally into two categories: crystals and glasses. The distinction is clearest on the nanoscale. The atoms in a crystal are arranged periodically, meaning they are spaced at regular intervals like the crests of a wave or the squares on a chess board. The strontium atoms in Madhavan's image appear in such a structure. Conversely, any solid with a disordered arrangement of atoms is called a glass. The glass of wine bottles is an example, but there are many other glasses, such as obsidian and some ceramics.

The distinction between crystals and glasses comes to the fore in the perennial disagreement as to whether glass is a liquid. It really depends on the timescales one is discussing. One piece of evidence often cited in defence of glass-as-liquid is that old church windows are thicker at the bottom than the top, suggesting they are very slowly flowing down. In fact this is misleading: the historic production method of glass involved rolling it out while hot (and decidedly more liquid-like), which led to window panes being produced with a thick end. This end was generally placed at the bottom when the windows were put in, and so it is incorrect to take the shape of old windows as evidence of glass being a liquid. Nevertheless, glass does actually flow – just very, very slowly. But so do some solids: lead guttering noticeably sags within a few years. The question is, on what timescale does the flowing take place? It seems sensible to call lead a solid, so flow on the timescale of years is probably too slow to be a liquid. On the other hand, some cheeses flow on a timescale of minutes or even seconds.

I spoke to Dr Camille Scalliet, a research fellow in Cambridge and an expert on glass and glass-like substances, and asked her where the research community draws the line. She replied that if there is an appreciable flow on a timescale of a hundred seconds they would consider something to be a liquid, otherwise it would be a glass (or something even more solid). So there you have it! It's a bit like saying four grains of sand makes a pile. It's an undeniably precise statement, but quite an arbitrary one.

The process of learning about glass is much like the three-stage appreciation of magic. First, you enjoy the show: nature has produced a solid which flows! Second, a little later, you learn the technique behind the magic and rationalise it into a broader worldview: glass is a liquid, so of course it flows. What, you didn't know glass was a liquid? Phhh. It's easy to get stuck at this stage. But third, a little later still, if you're very lucky, you learn you were wrong to dismiss the magic so hastily. Glass is an amorphous solid, or a super-cooled liquid (meaning it is liquid below its freezing point), and there exists a world of classification-defying materials like this, hidden in plain sight by our attempts to categorise. The world is magical after all, and now you can appreciate the show with the insight of the professional magician.

So solids might be familiar, but they still have their secret ways. How about the other elements?

Liquids, represented by water in the classical elements, cannot resist shear stresses. Recall the magician slicing cards off the deck: a liquid deck would flow into a puddle. On the nanoscale, a liquid is disordered. Yet liquids are still dense like solids. Gases, represented by air in the classical elements, also cannot resist shear stresses, and additionally they lack the density of liquids. The density of a gas depends on the masses of the molecules that comprise it. This formed the basis for a spell of levitation I once saw performed by Professor Kari Dalnoki-Veress, who describes his branch of condensed matter as 'squishy physics'. Dalnoki-Veress made a paper boat, and placed it in an empty fish tank. As if by magic, the boat floated around the empty tank as if it were on water. In fact the fish tank was filled with xenon, an invisible gas that is heavier than air. The xenon was so dense that the paper boat was lighter than the xenon it displaced, and therefore floated by Archimedes' principle. Next, Dalnoki-Veress proceeded to cast a spell of transformation, making his voice unnaturally deep: after inhaling xenon your voice goes deeper, just as inhaling lighter-than-air helium causes your voice to go higher. He had to stand on his head to demonstrate this, because otherwise the heavier xenon would have sat in his lungs and suffocated him.

Plasma, represented by fire, is distinguished from a gas by being 'ionised', meaning that some of its atoms and molecules have gained electric charges to become ions. Fire is a good everyday example of a plasma: because it contains freely moving ions, it naturally conducts electricity. Another reasonably everyday (or at least several days per year) plasma is lightning. This is certainly an electrical conductor, although it would be a decidedly mad scientist who set out on a stormy windswept night to test that with a voltmeter.

It is tempting to ask *why* some gases have become plasmas. Since plasmas are relatively unfamiliar to us in our day-to-day lives, it strikes us that they need more explaining. In fact, plasma is the predominant state of condensed matter throughout the universe; stars are great balls of the stuff. In essence, though, plasmas occur at higher energies, as the atoms become so energetic they lose some of their electrons. As a general rule, the progression of matter from low energy to high is earth, water, air, fire. You can think of it in terms of an argument of wizards (for that is the collective noun): at low energy they are all seated, solemnly discussing important matters, still like the atoms in a solid. But a rumour begins spreading about a new spell that has been learnt; they begin walking around muttering to themselves, moving about like the atoms in a liquid. The rumour develops into a story that this spell is of great importance; they begin jumping up and down with excitement and running around telling one another and asking what the spell might be; their rapid movement resembles the atoms in a gas. Finally, the rumour is confirmed: the spell is the long-sought-after spell to find one's wizard hat when it has gone missing, perhaps the most sought spell in all of wizardry; the wizards become frantic, dashing around like the atoms in a plasma, throwing their hats in the air and losing them.

Which classical element represents metals? Well, most metals are solid under ambient conditions, suggesting earth. But they also conduct electricity, suggesting fire. So maybe metal is a combination of earth and fire. Translating from ancient to modern terminology, that's pretty much right. Metals are solid, but in order to bond together their atoms have given up one or more electrons to become positively charged ions sat still within a plasma sea of negative charge which is free to move. So plasma is more common than it first appears, and doesn't require huge energies or high temperatures.

Collectively, then, the four classical elements do a surprisingly good job of matching the states of matter we see around us – the collective behaviour which emerges when many atoms interact. But there are many more than four states, some of which are rather familiar.

The fifth element

The four-states-of-matter idea has done well to survive for millennia. But it has also limited our thinking somewhat, blinding us to other states that we meet even in everyday settings. Liquid crystals, as used in laptop and television screens, are one example. Their molecules line up, which is not true of liquids, but they are not solids either: their molecules do not form an ordered pattern as they do in crystals, and they flow too fast to be glasses. Gels (such as jelly) and colloids (such as milk) have different properties from any of the four classical states. A colloid is a suspension of solid blobs within a liquid: in milk, globules of fat are suspended in water.

But perhaps the most familiar phenomenon evading classification in the four elements is magnetism.

A magnet does everything you'd expect a state of matter to do: the magnetic field emerges from the collective behaviour of many interacting atoms. There are magnetic solids, liquids and gases. Plasmas are inherently magnetic: a common design of nuclear fusion reactor, the tokamak, confines plasma using magnetic fields. Nuclear fusion involves the fusing together of atomic nuclei to release energy, as in the sun. It has no harmful by-products, cannot suffer melt-downs and only requires hydrogen as fuel; hydrogen is so abundant that fusion is considered a renewable energy. Fusion is yet to be em-ployed commercially; part of the reason is that it's rather difficult: holding plasma with a magnetic field is sometimes compared to bal-ancing jelly on a wire. The work of plasma physicists, who may or may not consider themselves condensed matter physicists depending on which of them you ask, includes developing the equations that enable this miraculous balancing act.

While familiar, magnets are mysterious enough to retain their magic. If you saw an object that looked to be moving around a table by the power of someone's mind, your first thought would probably be that it was a trick performed with magnets. And to be fair, it is pretty mag-ical that we can move things at a distance with magnets! Humans' first hands-on experience with them was with 'lodestones', naturally mag-netised minerals. No one is quite sure how lodestones came to be mag-netised; the leading hypothesis is lightning strikes, an idea supported by the fact that they only seem to occur close to the Earth's surface.

Lodestones grant another primordial glimpse of condensed matter physics, and they have always held an association with magic. One of the earliest extant references occurs in the *Guiguzi* (Book of the Devil Valley Master), from fourth-century BCE China. It reads:

Know the self and afterwards know the other. This mutual knowing is like the flounder, which survives only in a pair, and appears as light and shadow. Its investigation of what is said does not go amiss. It is like a lodestone drawing a needle, or a tongue pulling meat from roasted bones.

This was translated from classical Chinese by my friend Helena Laughton and my former student Sixuan Chen (now an experimental condensed matter physicist). Sixuan notes that pairs of flounder are a common metaphor for loving couples in Chinese poetry as they follow one another closely; traditionally they were thought to have only one eye each, needing to pair up to be able to function. The Chinese characters for lodestone literally read 'magnetised stone'.

Modern scholars believe the *Guiguzi* to be a collection of ideas from different authors. Historically the Master was believed to be a real person, and Devil (or Ghost) Valley his residence, long lost to the mists of time. In the second century BCE there are references to using lodestones to construct a magical spoon that would always point south when placed on a smooth surface: what we would now call a compass. The very word lodestone derives from the Middle English 'lode' meaning 'way' or 'route'. Traces of this usage still exist: my favourite pub is the Lower Lode Inn, a fifteenth-century tavern on the River Severn. The easiest way to get to it is by following a path south from the medieval town of Tewkesbury. You'll find yourself facing the pub across the vast expanse of the Severn. Look to your right, and you'll see a bell hanging from a post; clanging it will summon a ferryman who will cross the river to collect you. The alternative is a much longer walk across fields, or a much, much longer drive. Your best bet is to take the lower lode.

Lodestones are discussed at length in the 1558 book *Magia Naturalis* (Natural Magick) by Giambattista Della Porta. Known as the 'Professor of Secrets', Della Porta was a Neapolitan polymath whose expertise included cryptography, optics, astronomy, meteorology, physiology and playwriting. He founded the world's first scientific society, the *Academia Secretorum Naturae* (Academy of the Secrets of Nature). To join, applicants had to reveal at least one new secret of nature. This remains a condition for achieving a PhD in a scientific subject to this day: in modern terms, it just means making an original scientific discovery.

The Academy was disbanded by the pope under suspicion of sorcery; Della Porta was questioned by the Inquisition and many of his friends were imprisoned. Undeterred, he invented a method of passing secret messages to them – by writing on the inside of boiled eggs.* If the four elements grant primordial glimpses of condensed matter physics, *Magia Naturalis* surely grants a primordial glimpse of popular science. Della Porta lists swathes of ancient beliefs across twenty subjects ranging from 'Strange Glasses' to 'Counterfeiting Glorious Stones'. Rather than simply reporting the beliefs as fact, he documents his own experiments to establish their veracity and bluntly states when the ancients were talking nonsense. In Book VII, 'Of the Wonders of the Load-Stone', we find self-explanatory titles such as 'Chapter LIII: It Is False, that the Diamond Does Hinder Loadstones Virtue' and 'Chapter LIV: Goats Blood Does Not Free the Loadstone from the Enchantment of the Diamond'. He thoroughly cites his sources and many of his original observations are demonstrably accurate.

Now, while goats' blood has no effect on the power of a magnet, other magnets do have an influence. The form of the effect depends on the type of magnet. The most obvious magnets are those materials that are magnetic in and of themselves: ferromagnets. The prefix 'ferro' refers to iron, which is a typical example: all pure iron can be made to hold a magnetic field by training it with another magnet. Then there are materials that are not magnetic by themselves, but become magnetic when placed next to a magnet, to which they are then attracted. Remove the magnetic field and their magnetism disappears. These are paramagnets; the category includes most of the elements of the periodic table.

---

* Book XVI, Chapter IV of *Magia Naturalis*, 'How You May Write in an Egg', details no fewer than six methods. I admit I had no luck in reproducing Della Porta's results; the only method I can find reproduced on the internet is cutting a small hole and inserting a note.

Diamagnets are the third type of commonly occurring magnet. These are again not magnetic by themselves, but become magnetic when placed in the field of another magnet. But whereas a paramagnet is attracted to another magnet, a diamagnet is repelled. If you've ever placed magnets on things, you'll probably have found examples of ferromagnets and paramagnets, but you would be unlikely to have noticed any diamagnets. Yet they are the most common type – it's just that diamagnetism is usually too weak to notice. Diamagnets include water, wood, many metals, many plastics and most organic materials; you are diamagnetic, for example.

Diamagnetism can be put to some rather cunning magical uses. A magnetic field that changes sufficiently quickly across space is able to levitate any object using diamagnetism. There are practical applications of this: for example, mice can be levitated to simulate zero gravity without leaving Earth. Diamagnetic levitation provided me with one of my clearest personal experiences of the three-stage appreciation of nature's magic. It occurred when I visited the University of St Andrews in Scotland. Entering the office of some PhD students, I saw a small, thin crystal hovering in the air above a shelf. My first stage of appreciation was the obvious one: it was a levitating crystal! I could see no strings, no tricks of the light, no mirrors concealing supports. I poked the crystal and it moved, still levitating. I was convinced it was levitating by magnetism; by now I was on my way to the second stage of appreciating the magic: rationalising it into my existing conceptions. But here I faced a problem, because there is a mathematical proof called Earnshaw's theorem which says that it is not possible to levitate magnets in a static configuration. You can try this yourself: you'll never be able to balance a magnet in the air above others without using supports. I wondered if the crystal might be very slowly rotating, which would be one way around this. I asked the students, and they replied that it had been hovering there for months. That ruled out rotation, since air resistance would gradually have stopped it. Incidentally, the students didn't think to point the levitating crystal out to me, precisely because it *had* been there for months; it was familiar to them. But it was new to me, and its magic increased the more I failed to understand it.

In the end they revealed that the crystal was an incredibly strong diamagnet. It was a piece of pyrolytic graphite – which, under every-day conditions, is the strongest-known diamagnet in the world – and it was hovering above a bed of incredibly strong neodymium ferro-magnets. Earnshaw's theorem only rules out magnetic levitation by static configurations of *ferromagnets*, but this was a diamagnet. So then I moved to stage three; I understood how the crystal was levi-tating but could still enjoy the technical skill of the performance, an incredibly rare example of diamagnets and ferromagnets so strong as to cause permanent levitation without an energy supply.

All types of magnet emerge from the magnetic fields of individ-ual atoms – the atoms' spins. The name 'spin' is suggestive: spin-ning an electrically charged object would generate a magnetic field around it. The kind of spin we are talking about here is quantum mechanical, and everyday analogies are hard to come by. But as a basic picture to hold in your head, you can think of the atom's spin as coming from the negatively charged electron orbiting the nucleus like the Moon orbits the Earth. A circling electric charge generates a magnetic field (this is the basis of electromagnets, which pass electric currents along coiled wires to generate magnetic fields).

Of the three types of magnetism, ferromagnetism is special in that it is purely emergent: it only comes about because of the mag-netic interactions between many spins, which cause them to align. Paramagnetism and diamagnetism, on the other hand, can be un-derstood by modelling each of their atoms' spins as behaving in-dependently; their overall behaviours are nothing more than the sums of their parts. For this reason, some people argue that ferro-magnets are the only one of the three to be true states of matter, but I personally think this is too restrictive. One reason for thinking so has to do with a close relationship between the way that ferro-magnets turn into paramagnets on the one hand, and the way that water turns into gas on the other. Such changes are as important as the states themselves.

Phase transitions

The closest equivalent in ancient Chinese thought to Empedocles' fourfold division of nature is the *wuxing*, which identifies five elements: wood, fire, earth, metal and water. At the time of the *wuxing*'s inception there were five known planets, and the number five appears in many classifications. There were five directions (north, south, east, west, centre), five cardinal colours (black, red, jade, white, yellow), a type of tea for each colour, and five notes in a pentatonic musical scale.* *Wuxing* more accurately translates as 'five movements', and originally referred to the planets' wanderings across the fixed pattern of stars. The idea of attributing as much significance to changes between states as to the states themselves is important. The changes between states of matter are called 'phase transitions'. Phases of matter are a slightly more specific designation than states. Ice, for example, can have multiple atomic-scale structures. These are different phases, all of which are in the solid state. There are currently eighteen known crystalline phases of ice and one

---

* This is similar to the connected worldview espoused by Western alchemists such as Sir Isaac Newton, although in his time there were seven known planets rather than five. Newton decided there are seven colours in the rainbow because there were seven major notes in a musical octave, which related to the seven planets. We still get taught that there are seven colours in the rainbow, which is quite bizarre since a quick glance reveals a continuous spectrum.

amorphous phase. Both states and phases are connected by phase transitions. Let's look at these in more detail, focusing on the case of water and how it changes into steam.

Water may be familiar, but it's still quite magical, exhibiting many phenomena we have yet to explain. Sonoluminescence is an effect in which light is given off by tiny air bubbles in water when they are made to collapse by sound waves. There is no firm consensus on how or why this happens. Or take the Mpemba effect. When the physicist Denis Osborne visited the school of thirteen-year-old Erasto Mpemba in Tanganyika in the 1960s, Mpemba asked why water heated to 100°C freezes faster than an equal volume of water at 35°C when both are placed in the freezer. His schoolmates and teacher ridiculed him, but Osborne tested the idea experimentally and seemed to confirm the observation. The two published a paper on the phenomenon in 1969. There is again no consensus on how this works, and there is a debate as to whether it really does, or if the effect is a result of other factors not being properly controlled – for example, more of the hot water may evaporate so there is less to freeze.

Many of water's properties are quite bizarre. It is the only chemical that can exist as a solid, liquid and gas at ambient conditions. Unusually, when it is close to its freezing point, water is denser as a liquid than a solid, which is why ice floats. This strange property means the bottoms of lakes do not freeze, allowing fish to survive winter. If water did not have this property, we might not be around to wonder how strange it is. So there is subtlety in the solid-to-liquid transition.

There is subtlety, too, in the liquid-to-gas transition. Like water, air is both familiar and magical. (The gaseous form of water is not actually air, of course, it is steam: energetic water molecules *within* air.) Most of air's remarkable properties are shared to a greater or lesser extent by other gases, but since air is ever-present on Earth it is the gas that has been put to the most practical uses. Principal among its unusual abilities is its incredible thermal insulation: clothing keeps us warm by trapping air; the hairs on our body trap air close to us for insulation, which is how we can survive in the Arctic or sit in a sauna; the foam used to insulate houses works by trapping air. You'll have a first-hand experience of air's extreme power to insulate if you've ever held a lit match too long: you won't feel the flame until it is barely a millimetre from your finger. Dominique spotted a bit of air magic while she sat next to me as I was writing this book. She tried to warm her cup of tea on the stove, but noticed that the mug had a concave base which meant the stove could not efficiently warm the tea. But then we realised that this is taken advantage of in the design. All mugs have concave bases that trap air, insulating their contents from below (usually keeping them warm). This is why mugs leave rings, rather than disks, on coffee tables.* This power of air has been used most clearly in *aerogels*, human-made solids consisting almost entirely of air. A famous photograph depicts a flower lying on a millimetre or so of aerogel above a blue flame; the flower is unwilted. Aerogel is also incredibly light, transparent and able to support weight far in excess of its own (a 2-gram brick of it can comfortably support a thousand times its own weight). When used as house insulation, it can save masses of space and energy. By 2011 aerogel held fifteen Guinness World Records for its exceptional abilities; many of these are simply air's abilities put to use. Air, and other gases, derive these properties from their low density.

The gaseous form of any material is much less dense than the liquid form. The wizardly way of looking at this is to say that density serves as an example of an 'order parameter': some property

---

* There are other reasons for coffee mugs to have concave bases. First, if the base were made flat it might get small imperfections poking out, which would stop it from balancing correctly. Second, the mug rests on the ring in the kiln, which is how the concave base is able to be glazed.

that changes significantly across a phase transition. There are two types of phase transition, which we can understand in terms of the behaviour of order parameters.

The first type is known as a first-order phase transition. It is characterised by an abrupt change in an order parameter. Say you're boiling some water. Once liquid water has been heated to its boiling point of 100°C, an extra amount of energy, called a 'latent heat', has to be provided to turn it into a gas. Remarkably, it takes about ten times more energy to turn 100°C liquid water into steam as it does to heat the water from room temperature to 100°C. If you're making a drink that does not require boiling water, such as coffee or green tea, you can save a lot of energy (and therefore money) by switching the kettle off at the right temperature rather than letting the water boil and cool. An intuitive way to see where this energy goes is to notice that your kettle stays relatively calm for most of the heating but shakes violently in the last few seconds. The shaking, which requires a lot of energy, is caused by the sudden appearance of large bubbles, as the liquid turns to gas and the density decreases abruptly.

A practical use of first-order phase transitions is in 'phase-change materials'. I recall being entranced as a child when I was given a handwarmer as a Christmas present. It was a pouch of gel with a small metal disc suspended in it. When I pressed the disc, the gel instantly solidified, becoming warm in the process. How was this possible? Had the manufacturers never heard of the law of conservation of energy? Where had the heat come from? I later learnt the trick. Energy had earlier been provided to turn the material from a solid to a liquid: you had to microwave the pack to melt it when you wanted to use it again. The solid was the lower-energy state, but there was an energetic barrier to overcome in order for the liquid to freeze. Popping the metal allows the phase transition to occur, releasing the latent heat. It does this by providing a surface on which the solid crystal can begin to grow. Phase-change materials are now used to facilitate renewable energies such as solar power, which are available in excess at some times of day but lacking at others. The excess can be stored as latent heat, which is then released as thermal energy when required.

The second type of transformation is known as a continuous or second-order phase transition. A good example occurs when a ferromagnet forms. At high temperatures, iron is a paramagnet: the spins of its individual atoms point in random directions; while they feel one another's magnetic fields, the high temperature means they have too much energy to line up and behave collectively. As iron is cooled below 768°C, however, a phase transition occurs, and the iron becomes a ferromagnet. On the nanoscale the spins align, and on the macroscale the iron becomes magnetised.* The magnetisation serves as an order parameter, just as density did for boiling water. The difference is that in the case of the magnet there is no latent heat involved. Rather than jumping abruptly, the magnetisation turns on continuously as temperature decreases. This continuous change of the order parameter lends the transition its name. It is intuitive to think of the two types of phase transition by analogy to a landscape: both describe a change in height, but whereas a continuous transition is like a smooth slope, a first-order transition is like a cliff face. If the pressure on water is decreased, its boiling point lowers. This was well known to nineteenth-century explorers, who used the boiling point of water to estimate their height above sea level. It is also well known to all wizards who have found themselves braving a mountain pass and in need of a cup of tea: black tea must be infused close to 100°C for the full taste to emerge, so a wizard knows that they must take

---

* Strictly speaking the spins align within magnetic 'domains' a micrometre to a millimetre in length; different domains magnetise in different directions unless a magnetic field is applied to align them.

green tea for mountain journeys, as it should be infused in cooler water. Conversely, increasing the pressure on water, which could be achieved by descending deep into a mine, increases its boiling point. If there is no mine handy, an easier way is to heat the water in a sealed container: the pressure of the water will increase with its temperature.

At a high enough pressure, however, something really remarkable happens: the distinction between liquid and gas completely disappears. The result is bizarre, and surprisingly useful.

Supercritical fluids

Imagine you seek to prepare a life-restoring elixir through byzantine means – casting a flame under your alembic to facilitate the infusion of dried *Camellia sinensis*. Or, if you prefer, boiling water for a cup of tea. When water boils in a kettle its density changes abruptly: this is the cliff face of a first-order phase transition. But as we amble along the cliff towards higher pressures, we find that the height of the cliff lowers: the difference in density between the liquid and gas lessens. Eventually the top of the cliff meets the sea and disappears (Figure 2 on the next page): the phase transition separating liquid from gas ceases to exist. Exactly at the meeting point, the phase transition between water and steam switches from first order to continuous, just as you can get smoothly from the sea to the top of a cliff only at the exact point where the cliff top meets the sea. The temperature and pressure at which this occurs is called the 'critical point'. The critical temperature of water is 373.9°C, and the critical pressure is 218 times atmospheric pressure.

Above the critical pressure and temperature there is no phase transition separating liquid and gas. The result is something entirely new: a 'supercritical fluid'. Fluid is a catch-all term meaning something that flows. Unlike a gas, a supercritical fluid cannot be made to

2: Cliff meeting the sea.

condense to a liquid by decreasing its temperature. Unlike a liquid, it cannot be made to boil to a gas by decreasing its pressure. These results were established by scientist and inventor Charles Cagniard de la Tour (1777–1859). When he rolled a flint ball in a sealed gun barrel half-filled with liquid, Cagniard heard a splash as the ball hit the liquid. But above a certain pressure, the sound abruptly vanished. The interface between liquid and gas had disappeared: all was supercritical fluid.

Supercritical fluids are not so familiar in our everyday world, but they are not uncommon. The compressed air used to fill (non-floating) balloons is a supercritical fluid when stored in its cylinder. The atmospheres of the gas giant planets such as Jupiter are composed of supercritical fluids. And deep oceanic vents called 'black smokers' release supercritical water whose warmth and mineral content supports entire ecosystems of creatures unknown in the surface world: these deep-sea creatures are the only forms of life on Earth that do not ultimately derive their energy from the sun.

Importantly to wizards, supercritical fluids also have practical applications. They tend to make better solvents than liquids or gases, and they are better than liquids at infusing into porous solids. Supercritical carbon dioxide is used in dry cleaning, where it efficiently dissolves dirt. The process of decaffeination of tea and coffee uses supercritical fluids to dissolve the caffeine while leaving the flavour intact. Supercritical water can be used to convert waste organic material into hydrogen fuel cells and other energy sources, and can be used for carbon capture. So they're very useful; but their existence also raises an intriguing question about the nature of matter.

Returning for a moment to the cliff in Figure 2: you can start high on the cliff, walk down the slope into the water and then swim back to below your starting place without jumping off. Similarly, starting from a liquid, you can find a sequence of temperature and pressure changes that results in a gas, without ever passing through a phase transition. So are liquids and gases really different states of matter at all?

Much like the question of how many grains make a pile, the answer you seek probably depends on your definition, not to mention the reason you were asking in the first place. At atmospheric pressure, water and steam certainly have very different properties, and in everyday life it is useful to refer to them as different states. It turns out there is a way to make this apparent distinction between liquid and gas precise.

Wool and water

The earliest surviving written prose in Great Britain is contained in *The Mabinogion*, a twelfth-century middle-Welsh transcription of earlier oral traditions. The tales include early versions of the legend of King Arthur, and the work was a major influence on J. R. R. Tolkien. The tale *Peredur son of Efrawg* contains the following account of Peredur's magical journey:

> *And he came towards a valley, through which ran a river; and the borders of the valley were wooded, and on each side of the river were level meadows. And on one side of the river he saw a flock of white sheep, and on the other a flock of black sheep. And whenever one of the white sheep bleated, one of the black sheep would cross over and become white; and when one of the black sheep bleated, one of the white sheep would cross over and become black.*
>
> Translation by Lady Charlotte Guest

I'm afraid I can't explain what's going on in that passage; those were simply more magical times. Nevertheless, Peredur's sheep can lead us to a more formal understanding of emergence and states of matter. Imagine a giant grid of square fields, stretching as far as the eye can see. Each field contains either a black sheep or a white sheep, and each sheep can bleat to its four neighbours. At each instant, choose a random sheep: it changes colour if doing so increases the number of neighbours it matches. Unlike Peredur's sheep, ours will stay in their fields. If the colour of each sheep is initially random, what happens to the colours over time?

Since there are initially as many black sheep as white, it's tempting to imagine this must remain the case. But actually, after a long enough time, the sheep will either all be black or all be white. This works as follows. In the original random pattern there will be clusters of colour: for example, there might be a square block of nine fields all containing black sheep. These sheep are less likely to change colour: the sheep in the middle has four black neighbours so is as happy as it can be, while each of these neighbours has three black neighbours so is also very happy. Similarly, these clusters tend to grow with time. A large-scale result emerges from tiny initial variations on the scale of individual sheep.

Here's a picture of how the sheep's colours change with time. Each pixel represents one sheep.

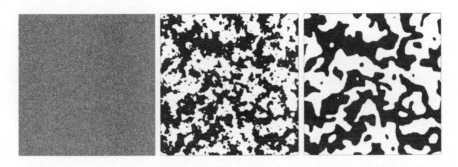

3: Peredur's sheep: starting from the left with random colours, with time they all become the same colour.

Peredur's sheep are acting out a classic model of magnetism known as the *Ising model* (pronounced ee-zing). The two colours of sheep represent two orientations of atomic spins. Ernst Ising was a Jewish German physicist born in Cologne in 1900. In his 1924 PhD thesis he found an exact solution to the problem that now bears his name. Fleeing the Nazis in 1939, he and his wife Johanna moved to Luxembourg – where he became a shepherd. Eventually they moved to the United States, where they lived long and happy lives: Ernst made it to 98, while Johanna narrowly missed out on her eleventy-first birthday, living to 110.

The fact that the Ising model can equally well describe magnets or magic sheep hints at its huge range of applications. This is the power of theoretical physics: the simpler and more abstract a model, the more phenomena it has a chance of capturing. The Ising model appears to many people in many guises. To computer scientists it is known as a Hopfield network and is a simple but powerful neural network for artificial intelligence. To biologists it appears as a simple model for memory formation in the brain. It has been put forward as a model for the behaviour of swimming bacteria, the movement of atoms in metallic alloys, and even the interactions of people when forming opinions. It is a running joke about condensed matter physicists that whenever they encounter a new phenomenon, their first thought is to use the Ising model to describe it.

Ising considered magnets whose atoms live along a line. Peredur's sheep instead recreate the Ising model on a square grid. This is one of the simplest physical models with a phase transition. Each spin must point either north-pole-up or north-pole-down, represented by the colour of the sheep. Each spin feels an attraction to its four nearest neighbours, making it favourable for them all to align.

With the precision of the Ising model it is possible to investigate the emergent states of matter and the phase transitions connecting them in more detail. One very clever way of identifying the emergent states is due to Leo Kadanoff (1937–2015), who taught me when I was a master's student at the Perimeter Institute for Theoretical Physics in Canada. He was an excellent and patient teacher. His knowledge of the ancient arts was unsurpassed, as he demonstrated by only providing a single course textbook – written in 1944. As you might expect from a person of austere tastes, his basic idea is elegant in its simplicity.

The Ising model is defined on the microscale. But suppose we'd like to know which states emerge on the macroscale. Essentially we want to blur our eyes and identify which features can still be seen. Kadanoff blocking, as it is called, is a way to formalise this mathematically. Let's think of it in terms of Peredur's sheep. Imagine you've spent some time studying under the wizard Merlyn in *The Once and Future King* and have the ability to transform yourself into a crow. Flying high above the fields you see that each sheep has been shedding its wool around its field, so that the whole field is the colour of the sheep it contains. Fly high enough that you can only just distinguish between individual fields. Now soar higher – so high that a block of nine fields appears the same size as an individual field appeared before. At this height, you can only resolve clumps of fields where a few sheep of the same colour appear together. Kadanoff's clever observation was that after this process of grouping fields together and zooming out, the result still takes the same basic form: it's still a load of black and white squares. Pick a set of spins in the Ising model and zoom out, and you get another set of spins in the Ising model. Repeat the process many times, soaring higher and higher, like blurring your eyes, and what remains is the emergent state. The form this state takes depends on the temperature of the magnet.

On the largest scale the spins would be happiest pointing all up or all down, because then every spin points in the same direction as its neighbours. This is a ferromagnet: all the spins align on the microscale to give an overall magnetisation on the macroscale. But real magnets experience the disordering effect of temperature, which causes spins to flip to unfavourable orientations. Peredur's sheep can account for this by not always changing colour when they'd like to, and sometimes changing colour when they wouldn't like to. At very high temperatures each spin will be randomly either up or down without concern for its neighbours. This is a paramagnet, with no overall magnetisation, and it is how the sheep start out: a random mix of black and white. So the scenario played out by Peredur's sheep is of a high-temperature paramagnet suddenly quenched to a low-temperature ferromagnet, like a blacksmith sticking a red-hot sword into a pail of water. At some intermediate temperature there must be a phase transition separating the two extremes of disorder and order.

The boundary between order and disorder is the phase transition: increase the temperature and the colours become a random mix of black and white; decrease the temperature and they begin grouping into blobs which grow as the temperature lowers. What happens in between, exactly at the phase transition? Is it random or blobby?

In fact, at the transition something remarkable happens: the pattern looks the same on all length scales. Take a look at the pattern below. The three pictures look like three variations on the same picture. But actually the middle picture is a zoomed-in version of the bottom left-hand corner of the left picture, while the right-hand picture is a zoomed-in version of the middle picture. Isn't that magic? What you are seeing is the behaviour of the Ising model at its critical point.

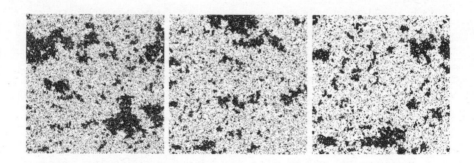

4: The Ising model at its critical point.

Given enough fields of sheep, you can zoom out as much as you like and it will look the same. Such patterns do not depend on the length scale at which they are observed (and their changes do not depend on the time scale on which they are observed). It reminds me a bit of the novel *The Third Policeman* by Flann O'Brien: locked in a police cell, the main character finds himself gazing at the cracks on the ceiling; he realises they form the same pattern as the roads in the town. Then he spots a moving cyclist: these *are* the roads in the town, viewed from far above; he must appear in miniature in this reproduction, looking at even smaller lines.

Yet reality is even more bizarre: the prisoner would have to zoom in by a fixed, discrete amount to see an accurate reproduction of himself, but the critical Ising pattern can be zoomed *any continuous amount* and will always look identical. Such patterns are said to be 'scale invariant', and they are surprisingly common, when you know how to look for them.

Scale invariance

Some familiar objects are approximately scale invariant. Clouds are a good example: from a distance they're white and fluffy, and look a little bit like the 'critical' pictures of Peredur's sheep; if you step into a cloud, in a fog or the misty mountains, it still looks white and fluffy; and if you keep looking closer and closer, zooming in, the cloud keeps looking essentially the same, right down to the level of a few tens of molecules. While your wooden staff is not scale invariant, the tree it came from has a reasonable scale invariance over a few different macroscopic length scales: trees have big branches branching off the trunk, small branches branching off the big branches and twigs branching off the small branches. Perhaps the best approximation to scale invariance is provided by the universe itself: galaxies exist in clusters, which live in clusters of clusters, and so on, with no length selected out as special.

Scale-invariant theories are accurate models for continuous phase transitions. Unfortunately, the phase transition you can most easily explore yourself – the boiling of water to steam – is first order and therefore not scale invariant. But the change of scale is itself instructive and something you can put to use. As you heat water and it begins to simmer, it emits a quiet high-pitched hiss as small bubbles form. As the water continues to heat, larger bubbles form and the pitch decreases. The sound turns from a shishhh to a shooshhh, which lets you hear and feel and see how close the water is to boiling. This is a useful trick when making different types of tea which require various temperatures to brew.

There is an ancient system of names for the different stages of heating water, dating back at least as far as *The Record of Tea*, written by a tea master called Cai Xiang in 1049–53 CE. The first stage is *shrimp eyes*, which occurs when the first, very small, bubbles form in the water, on the inside wall of the pan or cauldron. The water is about 70°C. The most delicate green teas will brew in shrimp eyes. The second stage is *crab eyes*, which are larger bubbles still attached to the sides. The transition from shrimp eyes to crab eyes is accompanied by steam beginning to rise from the water. The water is just under 80°C, and is suitable for most green, white and oolong teas. The third stage is *fish eyes*. The bubbles are larger, and are just beginning to release from the sides. This is when the water's song can begin to be heard. The water is just over 80°C, and some hardier green and white teas will release more flavour at this temperature. The fourth stage is *rope of pearls*, so called because the bubbles are now streaming to the top. The water is now at 90–95°C. This is the right temperature for black tea. The fifth and final stage is *raging torrent*, truly boiling water, at 100°C; the bubbles of raging torrent cause the water to lose too much of its oxygen which can cause some loss of flavour, although teabags are designed to require this temperature. The only loose leaf tea which can survive raging torrent is black ('ripe') *pu-erh*. This fantastic tea, which has a taste similar to earth (or goat washings, if you ask my friend Martin), is the most resilient, and can be made with any temperature from shrimp eyes up.

While it can be useful for brewing different types of tea, the fact that the size of the bubbles changes with temperature shows that boiling water is not scale invariant. However, if you heat water in a tightly sealed container you can increase both its pressure and temperature, allowing you to tune it to its critical point at which the bubbles become scale invariant. Bubbles form of all sizes, and you even get bubbles within bubbles, within bubbles, and so on, on all length scales from a few tens of nanometres all the way up to nearly the size of the container. All these bubbles scatter light, leading to a really magical effect: the water ceases to be transparent, and instead turns a milky white, resembling an opal. This is called 'critical opalescence'.

The theory of critical opalescence was developed by Albert Einstein as part of his work explaining how the middle realm of our existence emerges from the microscopic world of atoms and molecules. Einstein predicted that the scale invariance of the bubbles in water at its critical point would scatter different colours of light through different angles. In fact, his mathematical expression is exactly the same as that which describes the scattering of light in the atmosphere – the reason the sky is blue. This prediction was easy to confirm experimentally due to the striking appearance of critical opalescence.

Einstein's mathematical model accurately described the experimental observations, but it also did much more. A model isn't just some mathematical equation or an oracle to be consulted. A model is built on assumptions about the underlying physical reality. Evidence for the validity of the model provides evidence for the validity of the assumptions about the nature of reality, which are often inaccessible to experiment. Einstein's equations took the form they did because he assumed the macroscopic world emerges from a microscale world of atoms and molecules. While this is intuitive today, it was far from accepted in 1905: despite the concept of atoms dating back to at least the eighth century BCE (by the Hindu sage Aruni), by the late nineteenth century the competing idea that the universe was smooth and continuous had become widely adopted. When Einstein's model was precisely validated it lent convincing weight to the molecular theory itself. Without Einstein we might not believe in atoms to this day.

Ancient practitioners of natural magic sought universal laws which govern the behaviour of the world; modern physics achieved this aim with great success. Scale invariance at the critical point provides one of the clearest examples of universal behaviour. It's such a good example, in fact, that we call the phenomenon *universality*.

The universe in a grain of sand

In common usage the word 'universality' refers to the situation in which different things exhibit the same behaviour. There are many kinds of universality in physics. For example, in a 2011 paper in the *Proceedings of the National Academy of Sciences*,[2] scientists quantified some of the remarkable collective behaviours of fire ants. When water floods their nests they cling together to form balls several centimetres in diameter which allow the entire colony to float on water. When the ball finds land, the ants spread out and return to moving separately. The researchers quantified these collective ant behaviours in terms of solids and liquids; they measured properties such as the viscosity of the ant fluid, and even studied the solid-to-liquid phase transition in fire ant behaviour. In the context of phase transitions, universality is the observation that large-scale behaviour often does not depend on microscopic details. It was first noticed in experiments on the phase transition separating liquids and gases around their critical points: many fluids were found to have identical large-scale behaviours, despite being made from totally different types of atoms with totally different interactions. The critical points of these fluids occur at very different temperatures and pressures. Sitting at the critical pressure and cooling towards the critical temperature, the density of these diverse fluids changes in precisely the same way, in a manner that can be precisely quantified. The form this behaviour takes is bizarrely precise: specifically, the density of these fluids is given by the temperature raised to the power 0.326. There's nothing special about that number from a mathematical perspective, but that's exactly the point: the number's just a number, but totally different fluids are described by it. That's universality. And what's really weird is that exactly the same behaviour also governs *magnets* at their critical points! In this case it is the magnetisation, rather than the density, which changes with temperature; but again, it is governed by that same number, 0.326. The same behaviour can even be proven mathematically in the Ising model.

These days, universality has been identified in all manner of situations. Examples include the spread of cracks in metals and minerals; the ripping of paper; the soaking of water through filter paper; the spreading of molecules in solution; avalanches in sand piles; dropped connections on the internet; and the development of rigidity in growing embryos. In all of these cases universality refers to the appearance of the same seemingly arbitrary numbers in unrelated contexts.

Universality has even been found to govern human behaviour. The authors of a 2013 article in *Physical Review Letters*[3] analysed videos of mosh pits at rock concerts and showed that the emergent crowd behaviour matched the behaviour of the liquid–gas phase transition around its critical point. Each person moves according to the music and their immediate neighbours, yet 'circle pits' spontaneously form in the crowd, in which people move faster, with more collisions, giving the crowd a reduced density. The authors argue this is an emergent state of matter like any other. I have been in circle pits myself; I did not notice an obvious loss of free will when moshing to System of a Down, for example, even when they played 'Sugar' straight into 'Prison Song'. But perhaps I was like the swimmer in the whirlpool, going with the flow. The paper's authors note some practical applications of their work: panicked crowds fleeing fires show similar behaviours to mosh pits, which can lead to the crowds becoming jammed. The authors suggest their analysis be employed for safer architectural design and crowd-management strategies, noting, for example, that mosh pits have evolved a rule that when someone falls their neighbours pick them up. Why do these unrelated systems behave in the same way? Our modern understanding of universality is that it derives from scale invariance. Close to a critical point, physical systems look and behave similarly on all scales of length and time. This means that the microscopic details (atoms or spins or moshers) become irrelevant: they disappear when viewed with sufficiently blurred eyes. If a mathematical model is to capture such a situation, it must itself be scale invariant. From this perspective, a simple explanation for universality is that there are a limited number of possible scale-invariant theories, in much the same way that the scale-invariant pictures above are very special.

Universality and wizardry

What is matter? This chapter's answer is that it is the collective behaviour which emerges on the everyday scale when many microscopic components interact. Atoms and molecules combine to more than the sum of their parts, with the result being the states of matter we experience. The same set of atoms can form many different states, while many microscopic arrangements can be consistent with the same state of matter. Central to this is the idea of universality: small-scale details often become unimportant, with the same large-scale behaviour emerging in wildly different settings.

It's a bit like the Marvel Cinematic Universe: the small-scale details may change between films – maybe the protagonist is a wizard, or a scientist, or an alien god; maybe the piece of ill fortune that granted them superpowers was a bite from a radioactive spider, or an accident with some gamma rays, or a spaceship engine blowing up – but entering the cinema you can be confident that the story will be comfortably familiar, leading to a victorious climactic battle in the penultimate scene, before a light-hearted resolution. The details can change, but the large-scale behaviour is the same.

This chapter began with a simple idea of emergence: add enough grains of sand and a pile emerges. Similarly, adding together a wealth of theories and experimental observations leads to a more complex idea of emergence. For example, continuous phase transitions exhibit scale invariance, bubbles within bubbles on all scales of length and time. But true scale invariance is impossible: you can't have a bubble that is bigger than the container. Kadanoff had a simple way to handle this problem – phase transitions don't exist! Or rather, they only exist in mathematical models, which never quite match reality. This is a typical view among condensed matter theorists. But if the transitions between states don't exist, do the states themselves?

A pragmatic answer would be that states of matter exist to the extent that they are useful. The mathematical models are able to be precise because they can assume an infinite number of particles, which is not a real scenario. But you don't necessarily need your tea to survive such rigorous analysis: provided it sloshes about in your cup, as opposed to being frozen solid or flying away like a gas, it will serve its purpose. The answer also depends on the relevant scales of length and time. If a block of Camembert can survive through lunch, then it is sufficiently solid, even though such a brief moment would be too quick to catch the attention of an onlooking olive tree.

Universality is precisely why simple models are able to capture the essence of real physical situations. The complicated small-scale details become irrelevant, in a way that can be quantified. To borrow again from J. G. Frazer, universality is an order and uniformity: a secret spring which, when reached, opens up a boundless vista. There is a way to the world, sometimes hidden, which can be learnt through careful thought and observation. Magnets and condensation seem at first to be unrelated, but they are connected by deep roots.

A primordial glimpse of condensed matter physics represents the states of matter by the classical elements. This chapter focused on liquids and gases, as represented by water and air. Let us now descend deep into the Earth, focusing on its most magical of manifestations: crystals. Journeying through the many worlds of individual crystals will lead us to an altogether different understanding of matter; one which is all a question of symmetry.

III

The Magic of Crystals

Veryan descended the dusty rock steps of the spiral staircase, her path illuminated by the light of her stone. The steps were narrow and their surfaces uneven; each was smoothed with a dip in its centre from centuries of use. After what seemed like an eternity of identical doors, one on each level, Veryan reached the one she sought. Cutting the bolt, she opened the door to reveal a vast underground library.

The floor and ceiling consisted of cast-iron trellises with an intricate floral pattern repeating in identical three-yard-long squares. Through the trellises Veryan saw innumerable levels identical to her own. Over the aeons, the accumulation of magic within the books had seeped into the architecture. Some results were harmless curiosities: walking forwards three squares, right three squares, back three squares, and left three squares did not always return one to one's starting position. However, other effects were a more tangible concern.

When the magic first began to develop it had been welcome. A thick oak bookcase would slide to accommodate a reader, or twitch to suggest an overlooked manuscript. But over the ages, as the air grew thick with enchantment, the library slowly forgot the readers. The bookcases, once row upon row of perfect alignment, came to move erratically, sliding rapidly back and forth in unpredictable patterns. Some slipped to create new groupings of subjects beyond the comprehension of rational minds. Some huddled in pairs, interested in what one another's books had to say, while larger groups poured their manuscripts into indiscriminate pools between them. Eventually the readers abandoned the library to its madness. Yet abandonment does not imply disinterest: Veryan's presence in the library would already have raised an alarm, and she should consider herself pursued.

Stepping into the library, Veryan immediately rolled around a swiftly moving bookcase three times her height. The vastness of the library mirrored that of the task ahead; she would need to work quickly. Through years of study and careful thought Veryan had found common threads woven through tales of the library. Pulling at those threads, she had worked them in her mind into an understanding of its workings. The movements were like the vibrations of dew on a mist-covered spider's web, perpetually in motion without ever colliding. Each shelf felt the pull of all others, and pulled in turn. There were patterns within instants of time, changing but forever the same.

Racing and dodging through the stacks, Veryan arrived at the book whose promise of forgotten knowledge had driven her across uncountable leagues. Here stood a trestle table, in a region of eerie tranquillity surrounded by chaos. Upon the table lay a single large tome, two feet wide and several long, hide-bound vellum fastened to the table by a heavy chain. With no time to spare, Veryan began to read.

～

The magic of crystals

5: Drawings of ice crystals (snowflakes) by Olaus Magnus
(1490–1557).

A book of wizardry could hardly be complete without a discussion of crystals. They are a natural embodiment of magic: from the soft uneven dirt we pull these hard, dazzling gems, with flat faces and geometric edges and sharp facets, artefacts as ancient as the world itself. In their hearts lie colours never before glimpsed. There are transparent crystals, opaque crystals, translucent crystals which are milky, or dusty, or which are dull until the slightest turn reveals a flash of brilliance. They can light the dark, or fluoresce in nu-rave neon. They are a spell that casts itself, the remarkable tendency of order to conjure itself from disorder. How does the irregular earth give birth to such dazzling symmetrical forms?

The idea that order comes from chaos is an ancient one. The Greeks and Romans believed that the world was originally chaos; in Norse mythology Ymir, a primeval being, originally resided in chaos represented as a chasm. In ancient Mesopotamia the goddess Tiamet symbolised chaos and primordial creation; in ancient China this role was played by Hundun, a faceless being sometimes equated with a mythical creature called a Dijiang. The fourth-century BCE *Classic of Mountains and Seas*, a bestiary of mythical creatures said to have been described by shamans and sorcerers returning from their trips, describes Hundun as a four-winged, six-legged, faceless creature said to resemble 'a bag'. According to one commentary, the fact that a bag doesn't really resemble anything is kind of the point: primordial chaos is formless and blind. While order from chaos is a common mythical belief, its occurrence in real materials is no less striking.

Crystals are so important that we mark our species' development by the dates we put them to use. The Stone Age began around three million years ago when we began to knap tools from flint (composed of tiny quartz crystals). The Bronze Age began some time after 6000 BCE when we learnt to cast it: bronze, an alloy of copper and other metals (traditionally tin), is also a crystal. While it might sound strange that a metal can be a crystal, this is actually the rule rather than the exception: almost all metals are crystals. As our mastery of fire improved, we forged metals with higher melting points, entering the Iron Age in about 1200 BCE when we began crafting steel (a crystalline alloy of iron and carbon).

The atoms in metallic crystals each give up one or more electrons to form a negatively charged sea, a little like bookcases in an ancient magical library finding a mutual attraction by sharing their manuscripts. This makes metals good conductors of electricity and heat. The reason they are cold to the touch is that they carry warmth away from your hand. Evenly shared electrons mean that metals are happy to change their shapes, making them malleable (easily hammered into new forms without shattering) and ductile (easily drawn out into wires). Practical applications have been central to our relationship with crystals from the beginning, and learning their ways formed an important chapter in the primeval past of condensed matter physics. Nowadays we make constant use of their magic, from LED lights in our homes and streets to liquid crystal displays on phones and laptops to laser diodes used to send internet communication down fibre optic cables. Electronics are built from crystals in the form of silicon microchips.

My friend Stephen Blundell, Professor of Condensed Matter Physics at the University of Oxford, tells the story of crystals like this: each crystal is its own unique world with its own laws of physics: its own speed of sound, its own speed of light. In one crystal world, squeezing creates electricity. In another, charged particles move in circles. Let us set out on a journey between these worlds, in the fashion of Oxford-based miscreants such as Alice of *Through the Looking-Glass* fame, or Lyra of Philip Pullman's *His Dark Materials* trilogy. In some worlds we will befriend new particles; in others we will meet old friends in new guises. We will go to worlds where magical powers are commonplace and bring back some of these powers, learning from crystals how to see our familiar surroundings in a new and magical light. The thread we will follow through these worlds, the thread which binds them together, is symmetry.

Symmetry is the defining feature of crystals. As a child, I vividly recall my excitement at being offered a crystal of bismuth by the front cover of a magazine. It would be a welcome addition to 'The Peacock Museum' I operated out of a cardboard box.* Eagerly tearing open the plastic pouch on the magazine, I hoped to find a new special exhibit. I was devastated to find I'd been duped – the crystal was some kind of plastic fake. It had been too much to hope for a real crystal of my own. The forgers' mistake had been to make their fake too beautiful: metallic, it had an oil-like rainbow patina across its surface, and took the shape of a pyramid stepped like a Mayan temple. But it was too symmetric to have come from nature.

---

* The museum was primarily geological in nature, but also showcased select conkers, curios and *objets d'art* I had found littered about the field at the bottom of the garden. Entrance was free, but there was an admission charge of two pence to the special exhibition of the Peacock Ore which lent the museum its name. The largest, shiniest conkers were also on permanent display here.

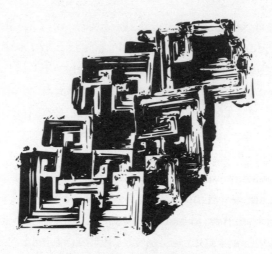

6: A bismuth crystal.

It was only years later, when I arrived as an undergraduate at Oxford and took a course in metallurgy, that I discovered the crystal had been real all along. Crystals really were magic: bismuth comes out of the ground with that unbelievable beauty and symmetry (Figure 6).

The importance of symmetry to the physicist's worldview cannot be overstated. To quote again from that master namer Philip Anderson:

It is only slightly overstating the case to say that physics is the study of symmetry.

It's a bold claim, but one whose defenders can be found throughout the subject. To understand why Anderson thought this we will first need to grasp the meaning and power of symmetry. Understanding how crystals achieve their symmetry is vital to understanding matter itself.

In this chapter we will meet the canonical definition of matter:

Matter is the rigid structure which emerges when symmetries spontaneously break.

To make sense of this statement we must journey through the crystal worlds, each with its own laws and peculiarities. But first, let us look at some of the magical powers of crystals, focusing on one that we can acquire ourselves to see the world in a new way.

Polarising opinions

Many crystals possess special powers. Magnetism is one example we have seen already. This is put to a fascinating use by magnetotactic bacteria, which have evolved to grow magnetic crystals within themselves in order to orient using the Earth's magnetic field, aiding their search for their ideal environments. The bacteria take in iron from the water which they use to grow either magnetite (iron oxide) or greigite (iron sulphide) crystals, from around thirty to one hundred nanometres in length. The bacteria have evolved to grow crystals of this size because these are large enough to give a significant pull from the geomagnetic field, but not so large that they break up into magnetic domains pointing in different directions (which would lower the overall field strength). Several research groups have conducted proof-of-principle work demonstrating that magnetotactic bacteria could be used to coat cancer-killing viruses, allowing the viruses to survive long enough inside the human body that they could be directed to tumours using magnets placed on the body.[1]

Another magical power possessed by crystals is triboluminescence, in which rubbing or banging together certain crystals causes them to light up. One of the earliest documented uses of this was by the Uncompahgre Ute people of North America, who collected quartz crystals from the mountains of Colorado and Utah and placed them in transparent shakers made of buffalo hide. When shaken, the crystals would bang together and light up. Triboluminescence is also exhibited by sugar: individual sugar grains are crystals, and you can see them light up with an eerie orange glow if you place them in a blender in the dark. Modern shamanic folklore has it that crystals of LSD can also be seen to light up when shaken in the dark.*

One of the clearest examples of a crystal's powers I've seen is birefringence. Calcite crystals have this power: when placed over the writing in a book, the crystal causes two images of the text to appear. Rotating it, one image waltzes around the other.

To understand how calcite works its magic on light it is first necessary to understand some lesser-known properties of light itself. Specifically, light has what is called a 'polarisation'. The basic idea, that light can be thought of as a wave, is not too tricky. You can create a wave along a rope by attaching one end to a post, standing back to pull the rope fairly taut, then waggling the other end up and down (Figure 7). You could equally well create the wave by waggling the rope left and right. Viewing the rope along its length, in the first case you would see that all the motion occurs in a vertical direction, while in the second case it occurs horizontally. The direction in which all the action occurs is called the plane of polarisation. If you could view a beam of light along its length as it travelled, you would see that it has an electric field wobbling up and down (say), and a magnetic field wobbling left and right. The direction in which the electric field wobbles is defined to be the light's plane of polarisation.

Depending on how light is generated, it may or may not be

---

* To establish this claim's veracity I consulted two masters of psychedelia: Danny Hammond supplements his shamanic work by performing as the lead guitarist in a psychedelic space rock band, while Dominique Scarpa has hosted radio shows and podcasts with a focus on the 1960s psychedelic movement. But our investigations were to little avail: the closest we got to a single crystal of LSD was a friend of a friend of a friend. Called Ulysses, he lives in a town beginning with B in either Austria or Switzerland, and was last heard planning to take the crystal to Goa. For now its triboluminescence must continue to reside in the liminal space between myth and science.

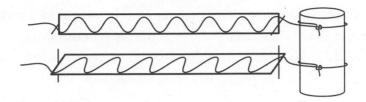

7: A wave made by waggling a rope attached to a post either vertically (top) or horizontally (bottom). This gives the polarisation of the wave.

polarised. Light from the sun is unpolarised, meaning you can find in it all angles of polarisation. Light from a liquid crystal display (LCD) such as a laptop screen, on the other hand, is heavily polarised. (Electric currents are used to align the molecules in the liquid crystals in the screen; light is only allowed through when the molecules have certain orientations.)

Within calcite, light with one polarisation will travel with one speed, while light polarised at right angles to this will travel at a different speed. When light slows down upon entering a material, it refracts and changes direction: this is why great skill is required to fish with a spear, as refraction causes the fish to appear in a different place from where it actually is. Since the two polarisations of light in calcite travel at different speeds, they refract so as to travel in different directions. If you shine unpolarised light into calcite, the beam splits into two (one for each polarisation), and you see two images of words through the crystal.

In our middle realm there is a fixed speed of light in the vacuum. It is the maximum possible speed and it is the same in whichever direction you look. If you were shrunk down to the quantum realm to enter the crystal world of calcite, though, you would find that there are two different speeds of light – and the speed depends on the direction in which you look.

Birefringence has led to the suggestion of a fascinating historical use for calcite crystals: they may have been used by the Vikings to navigate at sea. The suggestion is based on references in thirteenth- and fourteenth-century Icelandic texts to the use of 'solar stones' to identify the position of the sun in an overcast sky:

> *The weather was thick and snowy as Sigurður had predicted.*
> *Then the king summoned Sigurður and Dagur (Rauðúlfur's sons)*
> *to him. The king made people look out and they could nowhere*
> *see a clear sky. Then he asked Sigurður to tell where the sun was*
> *at that time. He gave a clear assertion. Then the king made them*
> *fetch the solar stone and held it up and saw where light radiated*
> *from the stone and thus directly verified Sigurður's prediction.*
>
> Rauðúlfs þáttr, c. twelfth century, translation
> byThorsteinn Vilhjalmsson

There are a couple of things to note about this passage. First, it seems pretty clear that Sigurður is performing some kind of magic, and it would seem reasonable to surmise that he is a wizard. Certainly the people around him seem to think the power is unnatural. If you could look at the cloud-covered sky and identify the location of the sun, wouldn't that be a magical power? Well catch that thought and bottle it, for this power will shortly be yours if you desire it. Second, note that the sun-stone does not seem to be treated as magical by the people in the account: it is the trusted method of obtaining the correct answer. To the Vikings, it seems the sun-stone is not magic – it is familiar. But then what is a sun-stone? It seems to have become forgotten lore, and in so doing has returned to the status of magic.

Vikings would have relied on cues such as the position of the sun or stars in the sky to navigate because magnetic compasses were not known in Europe until around 1300 CE. But near the Earth's poles, where Vikings were known to sail, such cues are often unavailable: for many months the far north lies in twilight, with neither sun nor stars visible. In a 2013 paper, scientists argued that a large crystal of calcite found in the wreck of an Elizabethan ship may have been used for navigation. The idea was this: the light of the sky is polarised, and the patterns in the polarisation paint a giant map across the heavens, which, if it could be seen, would identify the position of the sun.

Many animals, such as bees and ducks, are able to see the sky's polarisation and use it to navigate. The sky-map can be seen by humans with the aid of calcite. Rotating the crystal, the transmitted light turns from blue to yellow according to the direction of polarisation, picking out a contour on the sky map. Calcite was put to this practical use in the twentieth century by commercial pilots on polar flights; this was the basis of the claim that the mysterious sun-stone is calcite. Perhaps you don't foresee a major use for Viking navigation in your spellbook; that's OK, as birefringence has many other uses.

One of the most magical was revealed in 2011 when researchers in the United Kingdom and Denmark found that two calcite crystals can be stuck together to form a 'cave of invisibility'. Macroscopic objects placed in the cave are rendered invisible from the outside, from any angle, since light passes around them. Earlier attempts at invisibility cloaks relied on painstakingly building up designer materials from the atomic scale, and only worked for objects a few thousandths of a millimetre across and only for specific colours of light. The catch (for there is always a catch with practical magic) is that the calcite cave only works for light with a particular polarisation, so requires a polarised light source. To identify the presence of the cave, all a wizard would need to do would be to somehow learn to see the polarisation of light, like some kind of magical new sense. Surely, though, that lies beyond the realms of real magic ...

Second sight

There are two mammals known to possess the ability to see the polarisation of light without the need of a birefringent crystal. The first is bats, who use it to navigate. The second ... is humans. It is unknown why we have this ability, and most of us spend our lives unaware that we do. In 1844 an Austrian mineralogist called Wilhelm Karl von Haidinger was studying crystals under polarised light when he spotted a ghostly pattern resembling a four-leafed clover with leaves alternately yellow and blue. Sat in the centre of his vision, the leaves of the clover were about the width of his thumb at arm's length. What is truly remarkable is what happened next. When Haidinger removed the crystal, the image persisted. While the crystal enhanced the effect, Haidinger realised he could now see the pattern without its help. This pattern is now called Haidinger's brush. The two blue leaves lie in the plane of polarisation of the light.

Crystals had taught Haidinger he had a sense no other human had ever noticed. It is a sense that can be trained in any skilled wizard; as with all magic, however, the power comes at a cost, and you should consider the following carefully before deciding whether to learn it. The light coming from LCD screens is strongly polarised. If you hone your ability to see Haidinger's brush, you will forevermore see it in the middle of your field of vision when looking at your laptop or phone: a brown smudge which cannot be unseen. If this price is too high for you to pay, you should skip the next paragraph.

Ah, there you are. I knew you'd think any price worth paying for esoteric knowledge. Let's begin. Bring up a bright sky-blue page on your phone or laptop. If you rotate the screen back and forth quickly through a few degrees, you will see a very faint yellow-brown bow tie rotating with it wherever you look. The reason for the rotation is simply that it is easier to spot things when they are moving. At first the bow tie will just look like a brown smudge. Even fainter is a blue bow tie at right angles, which will look like a dark smudge on the blue screen. Once spotted, with practice you will be able to see the brush without rotating the screen. It tends to disappear after a couple of seconds of staring at a fixed spot, but moving your eyes will make it reappear. You have now trained yourself to see the polarisation of light; if someone has used a calcite cave of invisibility to conceal objects, you should now be able to see through the deception. The easiest way to see the polarisation map in the *sky* is on a cloudless day at twilight. Imagine a line passing through the sun which divides the sky perfectly into two halves. Point at the sun with both hands (without looking at it!), arms outstretched, then with your right hand trace along the line until your arms are at right angles to one another. Your right hand should now be pointing at the most polarised point in the sky, the easiest place to see Haidinger's brush. With practice you can see the map across most of the sky, and even through light clouds, like Sigurður on that thick and snowy day.

Haidinger's brush has an increasing number of practical uses. It has recently begun to be used as a test for age-related macular degeneration, the main cause of blindness in many parts of the world. The same parts of the eye that detect polarisation happen to be those that suffer the degeneration, so measuring a person's ability to see the brush provides a simple non-intrusive test. A related technique has been developed to identify visual impairment in young children non-intrusively, and training people to see Haidinger's brush has been shown to be effective in correcting a range of common vision issues relating to using the wrong part of the retina.

Much as only some animals can see polarisation, only some crystals are able to show birefringence, while other crystals have other powers. How do crystals work their magic? To answer, it is necessary to travel down to the microscopic world from which they emerge.

The crystal lattice

A crystal is a solid whose atoms are arranged into a regular structure; specifically, a periodic structure, meaning a pattern that repeats at even intervals, like a wave or railings or bathroom tiles. The atoms in a crystal are evenly spaced in all three directions, like the bookcases in the passage at the start of this chapter before they started going haywire. To see how this can be, imagine that the crystal is a warehouse like the one at the end of *Raiders of the Lost Ark*, filled with a huge number of identical cube-shaped boxes. The boxes are packed perfectly with no gaps between them (Figure 8, on page 76). Since a crystal is much smaller than a warehouse, the boxes will need to be tiny; imagine each contains one atom which fits snugly inside. Since the boxes lie in a periodic structure, so too will the atoms.

In a crystal, the atoms exist but the boxes don't. Rather, the atoms choose to sit where they do because of their interactions with the other atoms. Metals, we have already seen, have their atoms donate one or more electrons to a collective pool in which they then wallow; since the atoms become positively charged ions in this process, and the pool of electrons is negative, this sticks the atoms happily in place. Salt crystals, sodium chloride, instead exhibit ionic bonding, in which each sodium atom donates an electron to a chlorine atom so that they can both obtain more energetically favourable electronic configurations. The sodiums become positively charged and the chlorines negatively charged, causing them to again stick together. There are many other forms of chemical bonding which can occur.

Nevertheless, the regular arrangement of boxes is a convenient way to picture things. Physicists try to make perfect mathematical models which they hope share relevant characteristics with the imperfect world. The physics of crystals is underpinned by the mathematical idea of a 'crystal lattice'. You can think of this as like the non-existent boxes. The crystal lattice is an imagined, perfectly regular, perfectly repeating set of points (say, the centre of each box). Now, the real world is not perfect. Physical crystals do not go on forever in each direction like the lattice: even a fairly large crystal might only be a few centimetres in each direction. But there are a huge number of atoms in a crystal: something like Avogadro's number. For almost all of these atoms, on the microscale on which they experience the world, the surface of the crystal is so far away that the atoms do not notice its presence, so the approximation that the atoms repeat forever, perfectly arranged, is actually not so bad. The atoms in a crystal do not sit still; they vibrate back and forth (in a quantum way) about their favourite positions, in a process that can be viewed as the passage of phonons through the crystal, analogous to the movements of the bookcases Veryan was dodging in the library.

Any crystal can be completely described by two pieces of information: the set of atoms living in each box (identical between boxes), and the way the boxes are stacked to form the crystal lattice. Different lattices can be made by stacking differently shaped boxes. Some crystals, such as polonium, have cubic boxes. Others have cuboidal boxes (like cubes, but with rectangles for faces instead of squares). Topaz is an example. α-Quartz is made from boxes with hexagonal bases. Not every box shape is possible: only those shapes that can fit together with copies of themselves without leaving gaps. For example, you can pack hexagonal coins on a table without gaps, but you cannot pack pentagonal coins. In 1848 Auguste Bravais (a physicist whose other interests included the aurora borealis) proved that there are precisely fourteen different box shapes which can be used for crystal lattices. Every crystal in the universe has its atoms arranged into one of these fourteen patterns, now called Bravais lattices. There is no simple

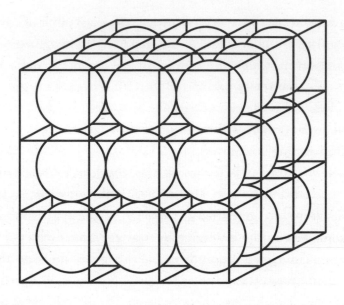

8: The atoms of a crystal; the boxes are a convenient abstraction.

reason why there are exactly fourteen. Fourteen is the number. No more, no less. Fourteen shall be the number thou shall count, and the number of the counting shall be fourteen. Sixteen is right out.

The Bravais lattices imbue crystals with their powers. Here, 'powers' really just means behaviours that would not occur were the crystal not there: differences between properties of the crystal world and properties of our own. We already saw that light can slow down and change direction in a crystal, and that it can take different speeds in different directions. There is no need for light to take the same speed in every direction in a crystal since, standing on an atom inside a crystal, different directions would look different. This is the origin of the birefringence of calcite. For a crystal to exhibit birefringence, light must behave differently in different directions, a property called 'optical anisotropy'.

Bravais lattices also explain the beauty of crystals. To return to the simplest example of cubic boxes with one atom per box: each atom in the crystal must find itself in an identical environment to all others. Whatever arrangement of neighbours works for one atom works for all the others, as they are identical. The crystal as a whole is built by stacking the boxes of the lattice. As a result the crystal approximates a giant version of those boxes. The flat faces of crystals are formed from perfect arrays of lined-up atoms. Sure, there are imperfections: missing atoms, extra atoms, the wrong types of atoms, and so on. Or there may be errors in the lattice itself, such as stacking faults in the boxes leading to remarkable effects in which stepping three boxes forward, three right, three back and three left does not return you to the original box. But by and large the story with crystals is 'as above, so below'.

The crystal lattice is the source of a crystal's powers. To understand how these powers arise, we must understand crystals' most striking feature: symmetry.

Fearful symmetry

In common speech if you said something was symmetric you'd probably mean it had a mirror symmetry: reflect it in a mirror and it looks the same. Is this the full moon I see before me, or its image in a still lake? This same intuition holds generally; we can say that

> an object has a symmetry when transforming it leaves it looking the same.

If an object has a mirror symmetry it looks the same when it is transformed by viewing it in a mirror. Many things look *almost* the same in a mirror, which has undoubtedly helped lend mirrors their otherworldly quality, enabling them to become staples of magical tales and horror stories. I recall being moderately terrified as a seven-year-old by a *Goosebumps* novel titled *Let's Get Invisible!*, in which a boy is slowly drawn into a mirror by his evil mirror-self. Since then mirrors have held a fascination for me. For example, have you ever wondered what colour a mirror is? Another mirror-riddle, popularised by Richard Feynman, caused me weeks of confusion as an undergraduate: why does a mirror reflect you left-to-right but not top-to-bottom? As a hint, for an object to have a mirror symmetry there must be a line through the object along which you can place a mirror so that the reflection perfectly replaces the missing part. I got a lot out of my weeks of confusion, but if you'd like my answer it's written here for your mirror-self.

The reason you interpret your mirror image as being flipped left-to-right is that you have an approximate mirror symmetry left-to-right (your left half looks like your right half in the mirror), but you don't have any such symmetry top-to-bottom (your head doesn't look like your feet in the mirror). To understand this, imagine instead a creature made of four squares – three blue and one red – arranged into a bigger square. The creature looks in the mirror. If it imagines the image to be itself after stepping around to the back of the mirror, it thinks the mirror flipped it left-to-right. But an equally valid interpretation is that it entered the mirror-world by climbing over the top of the mirror and dropping onto its head; in this case it thinks the mirror flipped it top-to-bottom. We don't imagine the latter scenario because we'd look totally different on our heads. In reality nothing is flipped: your left is on the left of the mirror. You just interpret it as a flip.

Mirror symmetries place important restrictions on the powers of crystals. For example, aside from a handful of processes undergone by certain elementary particles, our universe appears to possess a symmetry of handedness: as far as we can tell it is neither left-handed nor right-handed.* As my philosopher friend Professor James Ladyman once put it to me, it would be very strange if you built a mirror image of a car's engine and it was any less efficient. Bolt threads would run the other way (screws would be undone by turning them clockwise rather than anticlockwise), but so would the threads on the nuts. Anything which looks different from its mirror image would be flipped, but presumably the result would work just as well. However, travel into the crystal world of quartz and it's a different story. Quartz crystals grow in either left- or right-handed forms. This grants them the power of 'natural optical activity': when polarised light passes through quartz, the plane of polarisation rotates. To see this you can use a polarisation filter: a device that only lets through light with a given polarisation. Recalling the wobbling rope, you can think of the filter as a set of railings: the rope can wobble through the rails only if it's wobbling in the same direction as them. Say you filter the ingoing light so that you know it is all polarised in a vertical direction. Then, if you measure the light leaving the crystal, you will find that you must hold your second polarisation filter at an angle for the light to pass. The further the light travelled through the quartz, the more you must rotate the second filter.

---

* The weak nuclear force is known to be asymmetric with regard to handedness. There is a sense in which particles that feel this force, such as electrons, do not resemble their mirror images. But they do look the same under the combined process of flipping the sign of their charge, reflecting them in three perpendicular mirrors and having them travel backwards in time. The fact that this stronger symmetry is true for all laws of physics is known as the CPT theorem, where CPT stands for Change, Parity and Time reversal.

For a crystal to be optically active it must look different from its mirror image on the atomic scale. The atoms of a quartz crystal are arranged into either left- or right-handed structures, rotating the polarisation accordingly. Natural optical activity was discovered in 1811 by François Arago (a physicist, freemason and supporter of secret revolutionary societies – he gets a mention in *The Da Vinci Code*). It now underlies the operation of LCD screens, while also being used industrially to identify the sugar content of syrup: glucose and fructose molecules are mirror images of one another, rotating polarised light oppositely.

You might ask *why* calcite comes in left- and right-handed forms. The short answer is that it's possible for crystals to have such structures, and so they occur. If you think about it the other way around, it would be stranger if something were allowed by the laws of physics but never seemed to occur. Thinking back to the idea of the crystal lattice as stacked boxes, it is possible to perfectly stack boxes which look different from their mirror images. All the boxes must be the same (say, left-handed), meaning the resulting emergent structure has the same handedness as the microscopic components.

Not all the powers of crystals have to do with light. Quartz crystals also exhibit *piezoelectricity*: squeezing them generates an electrical potential – a voltage. If you think of the flow of electric current as like the flow of a river, a voltage is like a drop in height which pushes the current along. Piezoelectricity is used to generate the spark on some cookers and lighters, but it has many more applications: the global market for piezoelectric devices runs to tens of billions of dollars each year. One future use still under development is to place piezoelectric devices in the floors of public spaces such as train stations; in this way some of the energy expended by the movement of crowds can be converted into electricity.

The only crystals that exhibit piezoelectricity are those that lack 'inversion symmetry'. If a crystal has inversion symmetry, it would look the same if all its atoms were pulled through a point and out the opposite side, like a glove being pulled inside out (turning a left-handed glove into a right-handed one). Piezoelectricity derives from the fact that the molecules in the crystal have positively and negatively charged ends. Squashing the crystal changes their orientations and distributions, leading to an imbalance of charge. Inversion symmetry stops this because for every molecule that rotates one way there is another that goes exactly the opposite way so as to cancel the effect. This provides a tangible link from the microscale to the macroscale: squeezing a quartz crystal and seeing a spark jump off it immediately tells us that on its atomic scale it does not have inversion symmetry.

The fact that we can look at any box of the crystal lattice and see exactly the same environment defines a different type of symmetry: 'translational symmetry'. In maths and physics 'translation' means moving something without rotating or otherwise changing it. So a translational symmetry is present when you can move something along and the result looks the same. A full classification of crystal symmetries involves their behaviour under combinations of reflection, rotation, inversion and translation. When the possible symmetries of crystals were first enumerated in 1892 it turned out there are exactly 230 possibilities. These are called the 'space groups'. It seems strangely unsymmetrical that the number is 230: the space groups are an exhaustive list of every possible symmetry of every possible periodic pattern which can exist in our three-dimensional world. It seems like there should be a more satisfying number of them, such as three; but there's not.

While it is the microscale symmetries of crystals that lend them their powers, these symmetries often manifest on the macroscale. A familiar example (or not, depending on where you live) is provided by snowflakes. Each snowflake is an individual ice crystal. Since snowflakes have six legs, we can tell at a glance that the microscale structure of ice has a sixfold rotational symmetry. Isn't that neat? However, water's familiarity again masks its subtlety. For example, one question which confused me for a long time is this: why do all six legs of a snowflake look the same? The answer turns out to be an expert act of misdirection.

Letters from heaven

The magic of ice needs no introduction. The film *Frozen* was so popular it led to a 37 per cent increase in US tourism to Norway – and it wasn't even set there. Journeying down into the crystal world of ice reveals one of my favourite pieces of crystal magic.

Two inviolable laws of our universe are that the speed of light in a vacuum is constant, and that nothing can travel faster. But if you lived inside an ice crystal you would find the speed of light to be only about three-quarters of its usual speed. The speed is still constant, but a different constant. You know what's really cool, though (pun intended)? In this ice world other particles can now travel *faster* than light! There's no law against travelling faster than light in ice, only in a vacuum. Elementary particles called muons constantly rain down as cosmic rays (around 30 pass harmlessly through you each second) and routinely travel through ice faster than light. Just as a sonic boom occurs when the tip of a cracked whip exceeds the speed of sound in air, muons travelling through ice create a shockwave of blue light called Cherenkov radiation. This is put to use in the IceCube Neutrino Observatory, buried deep under the Antarctic. This experiment looks for ghostly elementary particles called neutrinos, which are notoriously difficult to detect. The experiment looks for the Cherenkov radiation bursts that occur when a muon is created by the interaction of a neutrino with the nucleus of a water molecule in the ice. The work at IceCube is central to dark matter searches because neutrinos are predicted to be measurable products of the decay of certain dark matter candidates.

I became interested in the symmetry of snowflakes when I saw them growing on the BBC programme *Frozen Planet*. How did one leg know what the others were up to? Asking around in the physics department, I was surprised to find that no one seemed to know. So I put together a project to have a master's student write a computer simulation to model snowflake growth. Now, the first place to start any research project is to see what's been done already. We found that the world's expert on snowflake growth is Professor Kenneth Libbrecht, former chair of the Department of Astrophysics at the California Institute of Technology. Credited as 'Snowflake Consultant' on *Frozen*, he also produced the video for *Frozen Planet* which had inspired me originally. Libbrecht's own interest in snowflakes arose during a visit to his frozen hometown in North Dakota. Struck by a desire to understand the origin of their beauty more deeply, he constructed a chamber in his garage for growing and filming them.

My student and I contacted Libbrecht. Aside from helping us develop our computer simulations, he also gave a simple answer to the question of why the six legs of a snowflake look the same. They grow within clouds, starting from tiny specks about half a millimetre across. The two things that affect the crystal's growth at each instant are the temperature and 'supersaturation' in the snowflake's immediate environment. When air is supersaturated with water it contains more water vapour than could exist in the presence of a solid surface. For example, if grass were around, the water would be deposited as dew. In a cloud the only solid things are the snowflakes themselves, so supersaturation is critical to snowflake growth.

As the snowflake whizzes around the cloud it experiences many different environments, and its growth changes instant to instant. Since no two snowflakes follow exactly the same route through the cloud, no two snowflakes look alike. But all six legs of a given snow-flake experience roughly the same environment at each instant and hence all look the same. But Libbrecht told us there was also more to the story.

The relevant factors in a snowflake's growth were first grasped by Ukichiro Nakaya (1900–1962), creator of the first artificial snow-flakes. When beginning his professorship of physics in Hokkaido, Nakaya found himself with limited equipment but plenty of snow, so it was to this that he turned his attention by creating the world's first laboratory-grown snow crystal (on the tip of a rabbit's hair). By carefully controlling the growth environment Nakaya devised what is now called the Nakaya diagram (Figure 9), documenting which type of snowflake will grow under which conditions of temperature and supersaturation.

Although there is not yet a general theory to explain everything in the diagram, certain trends are understood. At low supersatura-tion water is sparse, and a snowflake must wait for a molecule to arrive. This leads to flat faces, since the molecules prefer to stick where they will have the most neighbours, which occurs in the middle of faces rather than on edges or corners. At high supersaturation, all the water around the crystal will have been used up in growing the crystal; but if a bump on the surface can poke out through this depleted region it will find an abundant water supply, and will grow faster. This leads to the dendritic, fern-like growth, as smaller spikes branch off at all length scales.

Nakaya described snowflakes as 'letters sent from heaven': they contain a record of the whole sequence of conditions the snow-flake met along its journey.* Libbrecht went on to provide us with a

---

* Continuing his legacy, Nakaya's daughter, Fujika Nakaya, is an artist who, like her father, creates non-liquid water sculptures; but rather than ice, hers are sculptures of fog.

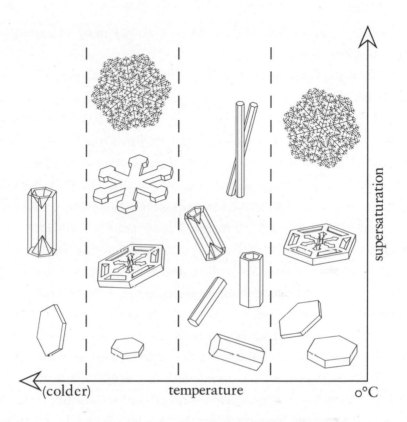

9: The Nakaya diagram of snowflake growth.

deeper reason why snowflakes are symmetric: they're not! It's a classic case of misdirection. For every snowflake image of his you see, he'll have discarded somewhere between 1,000 and 10,000 less-symmetric snowflakes. The myth of perfect snowflake symmetry was enhanced by the 1864 book *Cloud Crystals: A Snow-Flake Album* by Frances Chickering. To produce the book's images Chickering quickly cut out the shapes from paper while observing the snowflakes on her windowsill. To do so she invented the technique of first folding the paper into six sections. This saves time in the cutting by forcing perfect symmetry. I eventually managed to find some snow to investigate myself, and indeed found it hard to find any really symmetric examples. But I'd say they're all the more beautiful for their imperfection.

Chickering's exploitation of a snowflake's approximate symmetry highlights a major practical use of symmetry more generally: saving time by spotting patterns. By asserting that snowflakes are six-fold symmetrical, she only had to cut out the shape of one leg rather than six. This basic idea has a great deal of relevance to our modern world. Consider the data compression used to encode a video: the file tells the computer what colour each pixel must be at each instant. But the video would be way too big to handle if it contained the data for every pixel in every frame. Nearby pixels within each frame are often the same colour, which gives a form of symmetry to exploit: transform from one pixel to its neighbour and the colour remains unchanged. Similarly, most of the pixels stay the same colour between frames. That's also a form of symmetry: move to the next pixel and the colour stays the same. So one method of compressing video files is to record only the changes, relying on symmetry to fill in all other cases. It's a bit like writing 'chorus' in song lyrics: one word stands for many.

Snowflakes hold an important place in the histories of crystallography and mathematics. It was in his 1611 book *Strena Seu de Niue Sexangula* (The New Year's Gift or: On the Six-Angled Snowflake) that Johannes Kepler provided the first recorded explanation of why the macroscopic symmetries of crystals can be explained by their microscopic arrangements of atoms. Kepler proposed that the hexagonal shapes emerged from the tightest possible packing of identical spheres on the microscale, and he suggested that this packing must resemble layers of honeycombs (Figure 10). Remarkably, this conjecture was not proven until 1998. Kepler was inspired to think about the problem through his correspondence with English mathematician Thomas Harriot, who had himself been set the problem by the buccaneering bowls enthusiast Sir Walter Raleigh, who needed to know the most efficient way to stack cannonballs on his ship.

10: Kepler's drawing of the densest packing of spheres.

While Kepler's reasoning about ice was not entirely correct (water molecules are not spherical), his suggestion that crystals' symmetries originate in the microscopic arrangements of their atoms was far ahead of its time. Further support was added when it was noticed that the angles between the faces of all known crystals corresponded to the angles of those boxes that can be stacked. But as intuitive as Kepler's argument was, it seemed impossible to verify. How can you ever hope to inspect the arrangements of atoms? It was not until the twentieth century that a method was found: a method of travelling between worlds, from big to small and back again.

Through the looking-glass

Allowing X-rays to pass through you onto a photographic plate creates images of your bones, because the bones block the rays. But what happens when you shine X-rays through crystals?

The answer is an intriguing process called X-ray diffraction.

Diffraction is the effect that leads to the beautiful rainbows on soap bubbles and bismuth crystals and the wings of certain butterflies and beetles. When white light (a mix of all colours of visible light) hits a bubble, some of it reflects from the top surface of the film, and some from the bottom surface a tiny distance lower. When the light from the two surfaces combines, some colours are enhanced, and others are reduced. The effect is strongest when the wavelength, the distance between the successive peaks of the light wave, is about the same as the film's thickness: different wavelengths correspond to different colours, and so subtle changes in thickness lead to different colours diffracting differently, giving the rainbow.

Remarkably, you can control and detect the diffraction of light with no apparatus whatsoever. Simply press your thumb and forefinger together a few centimetres in front of your eye, with a clear light source behind them. Now separate them as little as possible, and you will see a dark bridge connecting them. Separate them a little more, and you will see the bridge divide into ten to twenty bands with light between, like the wooden planks of a rope bridge from *Indiana Jones and the Temple of Doom*. Your fingers are forming a narrow passage through which the light diffracts, as a water wave would when passing through a narrow strait.

I recall being similarly perplexed by some special 'rainbow rave' glasses I received in a party bag when I was about eight. When I put them on they showed rainbow-coloured replicas of everything I could see. How did the glasses know what I was looking at, in order to make copies of the images? Many years later I learnt the lenses were diffraction gratings: many tiny ridges scored so that light from different ridges added and subtracted like a kind of souped-up soap film.

One of the clearest demonstrations of diffraction can be found in the 'fire' of an opal – the brilliant flashes of coloured light you see as you turn the gem in your hand. Opals contain many tiny silica spheres, each about 1,000 atoms in diameter. These are arranged into sheets that reflect light like the surface of a soap film. The more tightly the sheets are stacked, the bluer the light that can be diffracted. A given opal might only be able to flash red, but if an opal can flash with violet fire (the shortest visible wavelength) it will also be able to flash with all colours. In his 1907 book *Precious Stones*, W. T. Fernie attests that opals, 'the most bewitching, most mysterious of all gems', were once thought to enact all the magic of other gemstones whose colours they contained: examples include the ruby, which loses its colour when danger is imminent; amethyst, the stone of temperance which restrains its possessor from indulging in too much alcohol; and emerald, which 'takes away foolish fears, as of devils, or hobgoblins'. Opal itself was believed to grant invisibility if held wrapped in a fresh bay leaf.

The diffraction of light from opals brings us to X-rays and crystals. X-rays are, in a sense, a form of light, although they are not visible to the human eye because their wavelengths are too short, residing far off the ultraviolet end of the spectrum. Narrow as their wavelengths are, the spaces between individual atoms in a crystal are even narrower, so when X-rays pass through crystals, they diffract.

The experimental process is simple enough: just shine X-rays through a crystal, then record the intensity at which they land on some photographic film (or a detector). The result is a set of regularly spaced tiny bright spots, like laser spots shone at the centres of every square on a chess board simultaneously. Figure 11 shows an X-ray diffraction image of a crystal of titanium diselenide provided by my experimental collaborator Professor Anshul Kogar at UCLA.

In the image you can see the set of sharp spots. You can also see some rings; Kogar explained that these are likely due to iodine

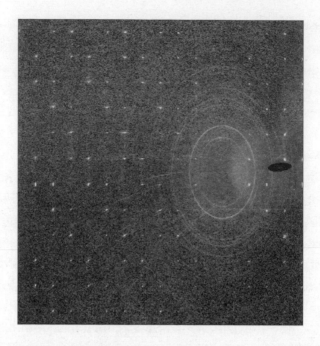

11: An X-ray diffraction image of titanium diselenide.
Courtesy of Anshul Kogar.

impurities. In general, rings are expected when crystalline order is absent: liquids give rings in X-ray diffraction, for example. In general, investigating the distribution of these spots reveals the arrangement of the atoms in the crystal. The twist is that the spots do not correspond directly to the positions of the atoms. Instead, short lengths in the diffraction pattern correspond to long lengths in the crystal and vice versa. The mathematical description of this small-to-big reversal is called a 'Fourier transform'.

Joseph Fourier (1768–1830) is famous for many advances in physics and mathematics, notably being the first person to propose the greenhouse effect of global heat retention. His was one of the seventy-two names inscribed on the Eiffel Tower when it was constructed. He is also notable for suffering one of the most ironic deaths on record. Having helped to develop the theory of heat, he convinced himself that staying warm was the secret to immortality. To this end he kept himself wrapped in blankets at all times. At the age of sixty-two he tripped over his blanket and fell down a flight of stairs, dying shortly thereafter.

Aside from providing us with this modern Aesop's fable, Fourier also provided the world with what is now known as the Fourier transform. It works a lot like a magical musical instrument. Imagine something like a piano, but whereas a piano has a fixed number of notes it can play, one for each key, this magical instrument allows you to play any note whatsoever. It's like a piano with an infinite number of keys (but let's imagine they all still fit into the same length – that's not so crazy, since there are an infinite number of fractions between 0 and 1). Well, that would be pretty magical already, but here's its true power: by holding down the right set of keys, pressed with the right firmness, this instrument can reproduce any sound imaginable. The hoot of an owl, the croak of a toad, the 1978 radio series *The Hitchhiker's Guide to the Galaxy*, anything. The notes must be held down continuously. The remarkable thing is, this instrument actually exists: your computer can act as such an instrument, for example. The reason it works is not the magic of the instrument itself: it is the bizarre fact that any sound whatsoever can be perfectly reproduced by the right mix of constant pure tones.

You will have had a hint as to how this works if you have ever tuned a stringed instrument: if you sound two strings which are close but not identical in pitch, you will hear a 'beat' as the overall sound gets louder and quieter rhythmically. By adding many pitches together more complicated sequences of sounds can be reproduced. The Fourier transform is the mathematical process of switching between the two things – the original sound that changes in time, and the set of tones each of which is individually unchanging. Now, time can be measured in seconds, while pitch is measured in cycles-per-second; more generally, the Fourier trans-form converts things with one type of unit to things with the recip-rocal of that unit (the reciprocal of a number is 1 divided by that number, so as the numbers get bigger the reciprocals get smaller). Intuitively, you may have noticed that big dogs have deep barks (with low frequencies better represented in the mix), while small dogs have high-pitched yaps: the bigger the dog, the smaller the typical tone of its woof, while tiny animals like mosquitoes make very high-pitched sounds.

The Fourier transform works with other types of wave as well. Imagine water waves adding up to give different shapes: long wave-lengths (equivalent to low pitches of sound) give the broad details of the shape, while short wavelengths give the fine detail. For exam-ple, imagine looking out over Loch Ness. You play another magical instrument, your thistle whistle, which local children have told you will summon the Loch Ness monster. In the distance you see a set of crests. Are they a wave with a wavelength of a metre or so, or are they Nessie's humps? Say you get a bit closer, and think you can discern some little ears on one of the waves. In order for them to be waves and not Nessie, there would have to be some short-wavelength ripples on top of the waves you'd already seen. A full, totally realistic profile of Nessie could be made given enough different waves.* Here's how it works with X-rays. A crystal lattice is nothing but a regularly spaced set of points. Its Fourier transform turns out also to be a regularly spaced set of points (the spots you see in the diffraction pattern). It is quite natural to think of both sets of points in a similar way: the regularly spaced atoms in the crystal are called the crystal lattice, and those in the diffraction pattern are called the 'reciprocal lattice'. To see

---

* OK, strictly, the classic Nessie profile has an air gap under its head and neck, which isn't possible. But any shape made from a set of humps can be made.

why, note that if you increase the spacing of the atoms in the crystal lattice, say by heating the crystal and causing it to expand, the points in the reciprocal lattice get closer together, in exactly the same way that when a number gets larger its reciprocal gets smaller.

When considering more general structures, physicists refer to 'real space' as the familiar place in which we live, and 'reciprocal space' as the world reached by the Fourier transform, where lengths and times transform into their reciprocals. Reciprocal space is a tricky concept, but the basic intuition that small goes to big and vice versa is the essential bit to remember.

To motivate the idea of it being thought of as its own space, consider again your piano that can play every possible note. You've heard a particularly good sound – say, a notably lonesome wolf howl – and would like to write some sheet music to tell your fellow wizards how to perform it. Since the instrument is played by holding down the keys continuously, it is tempting to think that all you need to specify is which keys to press, and how hard. That's pretty much right. We might imagine sheet music taking the following form: prop up a sheet of paper behind the keys, and draw a line on the sheet whose height above each key represents how hard that key should be pressed (the line drops to the bottom of the sheet when that key is not played at all). In this way, you can imagine shrinking in size and walking along the keys; if you want to know how a particular tone appears in the mix, you just walk to the relevant key, and measure how high the line is. The line resembles a mountain range in reciprocal space. If the wolf howl was deep, the mountains would be higher around the lower notes.

I first learnt about Fourier transforms and X-ray diffraction from Professor David Cockayne when I was a student in Oxford. Cockayne was Australian; his daughter was a flying doctor who had responsibility for an area of the outback larger than the United Kingdom.

The concept of reciprocal space is daunting at first. You have an in-tuitive understanding of how to move around in real space, because you have lived there your whole life. An area of the outback larger than the United Kingdom is a large area, and if you are called to an event within the area you can expect you will have a large distance to travel. In reciprocal space big becomes small, so flying around the reciprocal outback would be a much easier job than walking around your reciprocal garden. To get to grips with the idea, Cockayne in-structed us to read *Alice's Adventures in Wonderland* – specifically, Martin Gardner's *Annotated Alice*. Gardner was both a renowned populariser of mathematics and one of the most important magicians of the twentieth century. *Annotated Alice* contains footnotes that ex-plain the mathematical ideas hidden in Alice's magical world. The author of *Alice's Adventures in Wonderland*, Lewis Carroll, was an Oxford-based mathematician, and many of the fantastical episodes in the book contain mathematical allegories. (The sequel, *Through The Looking-Glass*, also contains the classic example of stepping into a mirror world.)

Cockayne instructed us to read the book in order to understand how to get to, and move around in, reciprocal space. Alice finds her-self unable to pass through a tiny door. She finds on a table a potion labelled 'Drink Me'. Upon drinking, she shrinks, a somewhat diffi-cult process, and once small can move around as normal. Cockayne explained this as an analogy for the Fourier transform: the process of transformation itself may be difficult, but, once in the small world (reciprocal space), moving around is as easy as moving around on the usual scale (real space). Just as Alice had to perform a different action to return to the big world, eating a cake labelled 'Eat Me', re-turning from reciprocal space to real space requires a different oper-ation: the inverse Fourier transform. However, the cake makes Alice bigger than she started out, which does not fit the analogy – there are only two spaces, real and reciprocal.

Which space you wish to exist in depends on the type of wizardly work you wish to conduct. As Alice found, some tasks are a lot easier in one space than the other. Here's an example: Professor Aude Oliva at MIT developed a magic picture which looks like Marilyn Monroe's face from far away, but Albert Einstein's face from close up. The picture works because it has only the coarse details of Monroe's face, and only the fine details of Einstein's. When I first saw it, I wondered how on earth it was made. Working with the pictures themselves, in real space, it's not at all obvious. But then it struck me – in reciprocal space it's easy: fine-grained features correspond to points far away from the centre, and coarse features correspond to points close to the centre. This is because small things go to big, and vice versa. So the process is this: Fourier transform each picture; take all the close-in points from Monroe's Fourier transform, and all the far-out points from Einstein's Fourier transform; combine the results in reciprocal space; and carry out the inverse Fourier transform to return to real space. Figure 12 shows my reproduction of Oliva's image:

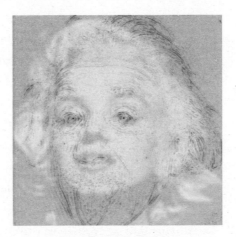

12: Marilyn Einstein: up close it's Einstein; from far away
it's Monroe.

The picture has valuable uses: when the image is flashed at people for a brief time they only see Monroe, demonstrating that the brain first identifies coarse features before later adding details. Oliva's group suggested constructing images that reveal different details when looked at from close up (say for layers of technical plans), animations which progress as you move the image closer to you, and text which can only be read close up, for privacy purposes. As a wizard you could easily work such an image into your repertoire, if you found yourself, say, needing to prosecute some unscrupulous monarch in a magical trial. 'Do you recognise this person, your majesty?' you might say, showing them a picture of a friend of theirs. 'What's that? You do?' [*Slowly turns image to distant jury to reveal the King's hired assassin.*] You could also use the picture to achieve a classic piece of stage magic: show a poor-sighted person some text close up and have them confirm that they can't read it. Then ask them to hold the paper up to the audience, who, being far away, will be able to read the text, seemingly confirming the person's poor sight. Then take the paper from them, say some magic words, and show them the text from further away, and they will miraculously be able to read it.

Fourier transforms are used all the time in the modern world. For example, they are another essential ingredient in data compression. In phone calls it is only necessary to transmit sounds between about 150 and 14,000 Hz as this roughly covers the range of sounds used in speech; other frequencies can be discarded. But how? First, Fourier transform the sound. Then throw away the fine (high-frequency) and coarse (low-frequency) data, just like with Marilyn Einstein. Then inverse Fourier transform back, and you've got a signal which can fit down a phone line. The same principle governs online streaming, for example.

The Fourier transform of a crystal is another crystal, but one that lives in reciprocal space. X-ray diffraction takes a photograph of this otherworldly counterpart, and in this reciprocal world, since small and large are switched, the tiny spacing of atoms is a help rather than a hindrance. X-ray diffraction granted access to Kepler's once-unreachable world of atoms and allowed us to verify that the symmetries of snowflakes grow from the symmetries of the crystal lattice. The photograph reveals the symmetries of the crystal's atomic arrangements, and with these symmetries lies the source of the crystal's powers.

The optical anisotropy in the calcite world – different behaviour of light in different directions – grants it the power of birefringence: two speeds of light varying with direction. The lack of mirror symmetry in the quartz world grants it natural optical activity: light's polarisation rotates. The lack of inversion symmetry grants quartz piezoelectricity: electricity causes space to contract. OK, really it's the crystal lattice that contracts, but if you lived down on the quantum scale within a crystal, the uniformly spaced atoms would be a bit like air is to us: ever present, and thought of as empty space.

While these effects are all bound to the symmetries of the crystal lattice, it is notable that they all derive from the *absence* of symmetries. There is a paradox in the heart of every crystal: while crystals are defined by their symmetries, they are really defined by the *absence of symmetries they might have had*. Crystals are what happens when certain symmetries break.

The mirror crack'd

Have you ever tried to balance an egg on its tip? There is a folk belief that it is impossible except at certain times of year, and egg-balancing festivals take place all over the world within that supposed window of possibility. In fact this myth was dispelled long ago by a physicist skilled in the art of symmetry. We've already met him: Ukichiro Nakaya, the inventor of artificial snowflakes, showed that egg balancing is equally difficult, but possible, at any time of year. The difficulty has to do with symmetry. An egg has a near-perfect rotational symmetry: turning it about its length, it looks the same. Nakaya explained that to balance the egg (Figure 13) you must find tiny imperfections so that three separate points of the shell simultaneously touch the table, with the egg's centre of mass sitting above the resulting triangle. If the egg were perfectly symmetric, the only way to do this would be to have the centre of mass sit precisely above the tip, which is not realistically going to happen. Knowing Nakaya's secret allows you to perform some magic of your own: sprinkle a few grains of salt on the table, and you will find it easy to balance the egg (three grains providing the three points of contact). Blow away the remaining grains, and it looks like the egg is balanced on the table itself. If you favour the dark arts you can readily work this into a profit-making wager in your local tavern.

A balanced egg is symmetric, looking the same from all directions. When the egg rolls, this symmetry is broken: the egg rolls in one direction only. How did it choose? Well, in reality, an egg is never perfectly symmetric. Nor is the table quite even: there might be a slight gust of wind, and so on. If we were to build a mathematical model we would be unable to account for such imperfections, and so we would just enact a bit of magic and say the egg chose randomly. We call this abstraction 'spontaneous symmetry breaking': a direction is chosen spontaneously. It is intuitive unless you get fooled by the perfection of your mathematical models. The fable of Buridan's ass warns us against precisely this mistake: an ass stands exactly halfway between two bales of hay, unable to choose, so starves to death in the middle. Crystals grow because they are able to choose.

The previous chapter considered liquid water and its phase transition to gaseous steam. Now let's consider its phase transition to crystalline ice. As we have just noted, freezing is an example of spontaneous symmetry breaking, because all crystals are defined by the

13: A balanced egg.

symmetries they lack. And yet, for that to be true, it must be the case that a liquid has more symmetry than a crystal. This might seem a bit suspicious – surely crystalline ice is more symmetrical than the random jumble of molecules that make up liquid water? But in fact, water is more symmetrical when it is wet. How can that be?

The solution to this puzzle hinges on the following point, which will sound like a trick but is a deep and mystical truth. Symmetries divide into two types: discrete and continuous. An equilateral triangle has a discrete rotational symmetry: rotate it through a third of a turn, a discrete amount, and it looks the same; but rotate it any smaller amount and it looks different. In contrast, a circle has a *continuous* rotational symmetry: rotate it by any amount and it looks the same. Continuous symmetries are thus stronger than discrete symmetries.

This turns out to be how liquids manage to be more symmetrical than crystals.

The defining feature of a crystal is that it has a discrete translational symmetry on the atomic scale: move the atoms by the fixed spacing between neighbouring identical atoms, a discrete amount, and it looks the same. Move them any less and it looks different. But a liquid has a stronger, continuous, translational symmetry, looking the same when translated through any amount, big or small. A liquid is disordered on the atomic scale, and it is equally disordered everywhere. Similarly, whereas crystals can have discrete rotational symmetries, liquids always have stronger, continuous, rotational symmetries.

Now, I agree this feels like a cheat – one conjuration too far – but I assure you it is not. The reason is that we do not live in the microscopic world of atoms – we live in our own middle realm. When we measure the properties of materials, whether with our hands or eyes or sophisticated experimental apparatus, we are measuring the average behaviour over a period of time and a region of space. On the atomic scale of a liquid there might be an atom here, or not, at a given instant; but the resolution of our measurements forces us to calculate an average over time, and on average the atom is as likely to be here as there. Similarly, the fact that our measurements have a smallest spatial resolution means that we are always considering quantities averaged over that smallest region. Provided this volume is large enough to contain a sizeable number of atoms, the liquid will look the same everywhere. It's these average quantities that are important to us.

X-ray diffraction makes this clear. Whereas diffracting X-rays through a crystal gives a grid of sharp spots (a photograph of the reciprocal lattice), diffracting X-rays through a liquid gives a ring whose radius is inversely proportional to the average spacing of the molecules (so that when the spacing gets larger, the ring gets smaller). The averages are the relevant thing we measure. So a disordered liquid is actually more symmetrical than an ordered crystal, and some of those symmetries must break upon freezing: it is worth noting that crystals break symmetry even on average.

Crystals derive their powers from broken symmetries. The more symmetries that are broken, the more powers the crystals can have (similar to the inevitable tragic backstory that lies behind a superhero's powers). Isotropic liquids look the same in all directions – a form of symmetry; when anisotropic crystals grow from these liquids they lose this symmetry, and in doing so they can gain the power of birefringence. Liquids have inversion symmetry; when uninvertible crystals grow from liquids they lose this symmetry, and gain the power of piezoelectricity. Liquids have continuous translational symmetry; when crystals grow from the liquid they lose this symmetry, with only a discrete translational symmetry remaining. All of these are instances of spontaneous symmetry breaking. The phrase 'spontaneous' indicates that of all the possible ways in which the symmetry could have broken, the crystal chose one option without being told how to choose. It has an atom here but not there – but what made 'here' better than 'there'?

Just as we invoked an unknown gust to explain how the egg rolled, we can invoke an unknown asymmetry to explain how the crystal grows. When water freezes to ice, the ice crystal will start growing from the wall of the container or from some impurity in the water. The atoms of the crystal are not equally happy in all positions, as they have to match up to the container. For a magical demonstration of this, if you cool extremely pure water very carefully, you can get it to remain a liquid several degrees centigrade below zero. In this condition, called a supercooled liquid, the tiniest knock will cause the water to solidify instantly. For the full effect on your audience, strike the container with your wand after uttering a suitably chosen incantation. This is the same process by which phase-change hand warmers work. A supercooled liquid is like a very hungry Buridan's ass, or a very symmetric egg waiting for the tiniest gust to knock it over.

The growing of crystals is inherently magical. Among physicists, crystal growers hold a uniquely wizardly position.

A large university department might have something like a hundred physicists, of whom most would be experimentalists and a handful would be theorists; but it would be lucky to have even a single crystal grower. Yet without them condensed matter physics could not exist. A single good crystal might take months to grow and will be passed around between the world's researchers for decades. They are lent freely, in return for authorship of a paper. My first publication in physics* came about when a crystal grower in Canada, Dr Harlyn Silverstein, managed to grow the first ever crystal of yttrium molybdate, which had been theorised to be a never-before-seen type of magnetic glass. In all likelihood this was the only crystal of yttrium molybdate that had ever existed in the history of the universe. It was priceless in the truest sense: there was no way to buy it, and nothing to compare it to for a valuation. I have a great respect for crystal growers. Twice now my enquiries about crystal growing have simply received the reply 'It's a dark art' followed by a pointed sip of tea to indicate the end of the discussion. All I can tell you is that when my friend Tom Brookes, an accomplished woodsman who can stick a hatchet in a tree at twenty paces, expressed a desire to become a physicist, and I suggested he become a crystal grower owing to his inherent magic, he presented me with a crystal of bismuth he'd grown in a pan over an open fire the week before. And when I went to Oxford's crystal grower, Professor Dharmalingam Prabhakaran, suggesting a crystal of cobalt silicide might contain certain never-before-seen quasi-particles, he replied that he'd grown a crystal of cobalt silicide the

---

* I say my first publication in physics, as my first published work was technically a theory I had on the nature of spontaneous human combustion published as a short comment in an article in the *Fortean Times*.

week before. I often ponder those tea-uttered words; I think their meaning is that beyond all the hard work and skill, growing that perfect crystal takes something more.

When Buridan's ass is between the bales the slightest hint would set it off to eat. But once that has happened, there is no getting the ass away from its chosen bale. The decision made, the ass sticks rigidly to its choice. And this rigidity turns out to be the key to a precise definition of matter.

Rigid thinking

Have you ever noticed how a wizard's favourite spells are always the most mundane? The novice always wants to do all the showiest magic, while the master barely seems to do any magic at all. Perhaps it is just another version of the three stages of appreciation: to the uninitiated any ability is impressive; the entered apprentice becomes familiar with the basics and wants something fancier; the master returns to the basics with the wisdom to see them in a new light.

Of the many arts devoted to human movement the same core principles always arise, with the most important being the most basic. Perhaps the most important and basic of all are breathing and standing. I once saw a clear demonstration of this in the form of a magical demonstration on a TV show. The presenter would pick a passer-by, ask them to stand still with their eyes closed, and, standing about ten feet behind them, he would seem to push and pull them with the power of his mind (and waving hands) until they fell over. It was quite a convincing deception, and I wondered how it worked. Then I recalled how difficult it can be to stand still: if you don't believe me just try standing on one leg with your hands by your sides and your eyes closed for a few seconds. So I tried it on a willing friend: I asked him to stand still, with his eyes closed, and said I would use magic to push him over from a distance. Sure enough, in a few seconds, he fell; even better, he claimed he felt a force pull him over. In a way he was right: the force was gravity. All the presenter was adding was a convincing set of hand movements to pretend he'd caused it.

If you ask me, the most magical property crystals possess is this same ability: they just stand there, holding their shape. Push one end of a crystal and the other end moves. Other states of matter cannot do this; for example, a liquid takes the shape of its container: push a liquid with your finger, and your finger pokes in. Recall that condensed matter physicists define solids as the state of matter which can resist shear stresses, that coordinated push–pull used to slide cards off a deck. In a liquid the individual atoms or molecules try to resist the shear individually, without coordination, and they fail. In the solid every atom is coordinated with every other: if you specify the position of any atom in the crystal, you know the positions of every other because they form a periodic structure described by the crystal lattice. Try to slide the top layer of atoms, and every other layer helps it resist. Crystals present a coordinated response by a macroscopic number of atoms that allows them to resist change. This is the definition of 'rigidity'. It is also a decent definition of matter itself.

What is matter? A definition you will often hear given by condensed matter physicists is this:

Matter is the rigid structure which emerges when the interactions between a huge number of particles lead to spontaneous symmetry breaking.

Let's break this down, using the example of ice growing from water. As water cools towards 0°C, the interactions begin to align the molecules into a regular periodic structure. What chooses where the molecules fix? Well it's the container, or imperfections, but in our mathematical model we say the choice is made spontaneously. Once the choice is made, each molecule wants to remain where it is, because its neighbours hold it in place, and their neighbours hold them in place, and so on. The structure is rigid.

Rigidity describes more than just crystals. Ferromagnets are another example we have already seen. Recall that each individual atom in a ferromagnet has a magnetic field called its spin, and that these spins align. At high temperatures the spins will point in random directions; all directions are equally unmagnetised, so there is a continuous rotational symmetry. Applying a magnetic field will readily turn the spins, as they respond individually to the force of the field. As the temperature lowers, interactions cause the spins to align, until at the phase transition they spontaneously choose a direction to point along. Once chosen, the spins stick rigidly to their choice: applying a magnetic field to try to rotate a spin, the spin will resist the force, because it would rather keep agreeing with all the other spins. The definition of matter as rigidity covers solids and ferromagnets, but I'd say it does not include everything we'd want to refer to as matter.

Of the four classical states, only solids meet this criterion. Yet there are plenty of condensed matter physicists who study liquids, gases, plasmas and myriad other things. This was the reason why Philip Anderson and Volker Heine renamed the subject from 'solid state physics'. Being less rigid in the definition of rigidity allows the inclusion of the other states. While liquids lack the shear rigidity of solids, they have rigidity in a more general sense: when astronauts pour water aboard the international space station, it doesn't just fly apart into individual molecules but chooses to stick together in a sphere. Poke it, and it doesn't fly apart. That's a coordinated response. This resistance to change can also be looked at as rigidity. The water molecules have condensed to form a state of matter. Even gases are condensed, choosing to stick together, albeit weakly: they are still described best in terms of their emergent behaviour rather than the behaviour of their individual molecules.

I should say that this more general sense of rigidity is not something all condensed matter physicists would agree on. Some might object that if a liquid or a gas were placed in the vacuum of space the atoms would fly apart. But that's true of solids as well, just on a longer time scale: when light bulbs contained a hot metal filament they had to be filled with an inert gas such as argon, otherwise the filament would evaporate and burn out too quickly for the bulb to be of use. I would say a set of atoms has this more general sense of rigidity if the interactions lead them to show some collective emergent behaviour.

Regardless of exactly how you define it, rigidity lies at the heart of condensed matter physics: the interactions of huge numbers of individual components lead to the emergence of a collective behaviour that is more than the sum of the parts. No single property of the individual atoms can explain how a crystal is able to stand, or why it has its symmetries. I have put forward a case that all matter can be defined by rigidity, and, by extension, symmetry. But what of Anderson's suggestion that all of *physics* is the study of symmetry? To understand this will require one final generalisation.

Possible worlds

For physics to be the study of symmetry, symmetry would need to be a guiding principle for the other branches of physics as well as for condensed matter. For some branches this is undoubtedly true. The standard model of particle physics is entirely built around symmetries. While a little more abstract than the symmetries of a crystal, those of particle physics are just as precise. And the same intuition holds: symmetries are when you change something and it looks the same. For example, change the charge of an electron to positive, and the result is the positron, which is otherwise identical. In the history of the universe as told by particle physics and cosmology, spontaneous symmetry breaking plays a starring role. In a process analogous to the growth of a crystal from a liquid, the Higgs boson breaks a symmetry of the universe to give the gift of mass to all those elementary particles that possess it. This is called the Anderson–Higgs mechanism, and it was first understood by Philip Anderson in the context of condensed matter physics before being applied more generally. But these are grand ideas far from the middle realm of mortals, and we will speak no further of such things. (Oh, except a bit in Chapter VIII.)

More generally there is a poetic sense in which physics is the study of symmetry. Physics seeks to rationalise the world and look for universal connections. The condensation of water at its critical point is described by the same mathematics as the development of magnetism. This is a form of symmetry: change water for magnets and the mathematical model looks the same. In this sense the physicist's search for hidden connections is the search for symmetry.

The perfectly flat faces of crystals, and the geometry of their facets, emerge as the everyday manifestation of their perfectly regular atomic order, as do their plethora of magical properties, from the obvious feats of magic such as glowing when shaken, right through to the everyday practical fact that pushing one end of a crystal makes the whole thing move. What's even more beautiful, to my mind, is that this regular periodic structure facilitates a whole world of emergent quasiparticles that can't exist outside crystals.

Some of these are entirely new. Phonons, particles of sound, cannot exist without a medium in which to travel. Others are old friends in new guises. Photons, particles of light, travel more slowly through crystals, and other particles can move faster than them. The result is the eerie blue light of Cherenkov radiation. This effect was originally predicted for elementary faster-than-light particles, before

Einstein explained they could not exist. That's the magic of condensed matter physics: an exciting phenomenon was predicted, it
was shown to be impossible in our world, but then it was discovered
to be possible, after all, in a different one. That world was discovered
within crystals.

Are there any limits to what can exist in these different worlds?
Our job as wizards is to spot hidden patterns within the chaos. But
how can we proceed, if anything goes? Fortunately, it appears that
there are rules that unite all possible worlds. Let us then proceed on
our journey, and examine at close quarters some of these laws that
bind all conceivable universes.

IV

Reflections on the Motive Power of Fire

The vast book open in front of her, Veryan began to learn the secret history of the world:

If you sail west of the westernmost point on the map you will find an archipelago whose islands constitute the most remote, and most magical, corner of the world. The climate is perpetually and pleasingly warm. Sufficient rain falls to sustain a comfortable and varied diet for the islands' inhabitants: fruits and vegetables, but also a certain root beloved to locals for its intoxicating quality. Every village, of which most islands have several, is home to a knot maker. On a day-to-day basis, the village knot maker might be responsible for such tasks as making the fishing nets, patching up sails, or tying together the bamboo struts of the beach huts. They might also, from time to time, be called upon to design a new knot, usually necessitated by the development of new technologies [windmills being a notable example]. The knowledge of new knots is shared amongst the islands by their creators; it may take many years to reach the more conventional regions of the globe. The design of a new knot often requires several rounds of discussion between the knot maker and other villagers in order to make it sufficiently understandable to catch on. For a knot to achieve that rare accolade of being considered truly beautiful it must combine practicality, necessity, aesthetic appeal and simplicity.

When knot makers convene, they can spend hours amusing one another with puzzle-knots containing fanciful loops and follies to confound the skilled practitioner. A favourite competition involves the picking of locks, which also falls under the auspices of the knot maker's art. Each knot maker begins with several interlinked locks, the aim being to unlock the bunch as quickly as possible; as a lock is picked, it is then relocked onto the rival's bunch ...

The limits of a wizard's powers

Wizards always have limits to their powers. Even if they are the most powerful being in their universe, there are usually restrictions set by the world they inhabit. For example, in *A Wizard of Earthsea* magic is limited by the requirement of knowing the true names of things.

So what are the laws of the world? And what can learning them tell us about matter? Alchemy sought to convert one substance into another, and to achieve immortality through the creation of an elixir of life. Alchemists are generally considered to have failed in this task, although alchemy led to chemistry, which can convert many substances to many others. To the extent that alchemy failed, it did so by not establishing the bounds of possibility: if you perform an allowed conversion of one substance to another, that's chemistry; if you try a forbidden conversion, it's alchemy and will fail. Copper into verdigris? Allowed. Lead into gold? Disallowed. Eggs into omelette? Allowed. Omelette into eggs? Disallowed.

Fortunately there are rules to the game even if we don't always know we're playing. In our tour through the classical elements we have examined water, air and earth. But fire is the element of trans-formation – for fire is energy, and fire creates chaos, and energy and chaos are the guiding principles that govern the world and its trans-formations. These principles are captured in the three-and-a-half laws of thermodynamics. These laws govern everything from cells to stars and all in between; but their origins lie in the practicalities of the middle realm. The laws began with Thomas Savary's 1698 patent for 'A new invention for raising of water and occasioning motion to all sorts of mill work by the impellent force of fire'. It was to lead to the development of the steam engine.

Now, I appreciate it might be difficult for most people to get too inspired by steam engines. Even steampunk often doesn't seem to involve much steam (or much punk), primarily limiting its focus to cogs, goggles and top hats. But the laws of thermodynamics apply to more than just steam engines, furnaces and pistons: they set limits on the behaviour of all matter, energy and information in the uni-verse, from DNA to black holes. Steam is just where it all began.

The first steam engines allowed us to bottle chaos. The result-ing Industrial Revolution was a Faustian pact: granting us mas-tery over matter and energy, our engines, like Faust's demon helper Mephistopheles, enacted our will in the microscopic world to effect changes in the macroscopic. This progenitor of condensed matter physics led scientists to their now-infamous belief at the end of the nineteenth century that they were approaching the end of knowledge.

But was Faust the master, or was the demon? As early as 1558 Giambattista Della Porta, in his *Magia Naturalis* (Natural Magick), observed:

There are two sorts of Magick; the one is infamous, and unhappy, because it has to do with foul Spirits, and consists of incantations and wicked curiosity [...] The other Magick is natural; which all excellent wise men do admit and embrace, and worship with great applause; neither is there any thing more highly esteemed, or better thought of, by men of learning.

It is fortunate, then, that we have elected to keep the latter but not the former. Demons are no longer widely believed in; arguably it was this exorcism of demons from otherwise rational beliefs that led to modern science.

In many ways alchemy gets a worse rap than it deserves. Sure, summoning demons is a bit old-fashioned. But aspects of alchemy are better thought of as early science. Many alchemical advances were real: the first recorded alchemist, 'Mary the Prophetess' (working in Alexandria around 200 CE), invented devices still used by chemists today, such as the bain-marie ('Mary's bath'). The influence of hermeticism (ancient occultism) in early chemistry survives in the name 'hermetic seal', referring to the airtight technology developed by alchemists. And many of the founders of the scientific revolution in the seventeenth century, notably Sir Isaac Newton, were self-avowed alchemists. When thermodynamics arrived in the eighteenth century it was driven by practical objectives, causing it to cast off the more esoteric parts of alchemy and creating modern science in the process. Only that which worked was kept. But that's not to say there was no space for theory: rather, thermodynamics survived because the theory was so successful. Through abstraction, thermodynamics achieved for itself the alchemists' dream of immortality.

In this chapter we will look at the laws of thermodynamics in turn. There is a memorable lay summary of the laws attributed to the beat poet Allen Ginsberg (although more likely originating with author and physicist C. P. Snow):

The zeroth law: you're playing a game.
The first law: you can't win.
The second law: you can't break even.
The third law: you can't stop playing.

(I must note in passing that these are also the rules of *The Game*, in which the game is to not think of *The Game* – which I'm afraid to say you just lost.) We will see the bounds these laws place on the types of magic we might perform, and how these laws emerge from the microscale. Historically the development of thermodynamics might be looked at as the immediate precursor to condensed matter physics; it led to a complementary view of matter as a balancing act between energy and chaos. Before looking at each of the laws, let's first visit the development of thermodynamics itself.

Carnot's waterwheel

In his 1824 work *Reflections on the Motive Power of Fire and on Machines Fitted to Develop that Power*, which lends this chapter its title, Sadi Carnot (1796–1832) derived the maximum possible efficiency of any engine that extracts useful motion from heat, such as a steam engine. This is often seen as the beginning of thermodynamics. From its inception the subject had infernal applications: Carnot sold steam power with the promise of victory in war and the expansion of empire. Fortunately Carnot's masterstroke lay in abstraction. A steam engine is a complicated apparatus of metal rods and pistons, coal, chimneys and human operators. But Carnot saw through to the essence of the process: he saw something universal.

As a child Carnot was fascinated by a mechanical waterwheel built by his father, powered by the flow of water from high to low. I can relate to this – in my hometown of Ottery St Mary in Devon there's a 'tumbling weir' in which a river drops into a giant hole. The weir once

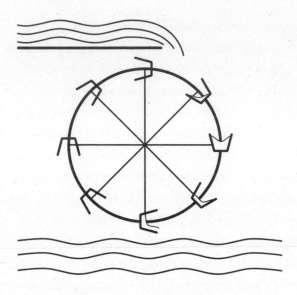

14: A waterwheel.

diverted water into a factory to turn a large wheel (still operational the decade before my granddad began working there in the 1950s).

To model a waterwheel we can imagine water flowing from a large reservoir, over a wheel, to a lower reservoir, creating the useful motion of the wheel along the way (Figure 14). Conversely, energy can be expended to turn the wheel backwards and return water from low to high. Carnot similarly imagined that heat flowed from a hot 'reservoir' to a cold one. He imagined the thermal equivalent of a reservoir to be a source of heat so large that it can give out heat without itself changing temperature. In the process of passing heat from hot to cold, useful motion can be extracted. In the steam engine, hot, pressurised steam is allowed to expand and cool, driving a piston. The hot reservoir is the source of hot steam, while the cold reservoir is the cooler steam released at the end.

Carnot deduced the maximum efficiency of any heat engine, now called the 'Carnot efficiency'. The greater the temperature difference between the reservoirs, the greater the efficiency. If you were limited to using liquid water in your heat engine, which can only exist between 0°C and 100°C, the Carnot efficiency would be only 27%. No amount of engineering would ever get you beyond that: it's a limit imposed by the universe.* Steam power might seem a thing of the past, indelibly linked to Victorians in goggles, or to steampunk fantasy worlds of dilapidated castles ambling around hillsides powered by friendly fire demons, like Calcipher from the Studio Ghibli film (and Diana Wynne Jones's book) *Howl's Moving Castle*. But actually steam remains thoroughly relevant. Around 85% of the power generated on Earth today uses steam. Nuclear power creates hot steam which, in expanding and cooling, drives a wheel (turbine). The same process governs the power plants of the future too. All of Iceland's mainland electricity is produced renewably, coming from the country's vast geothermal and hydropower supplies. The conversion of geothermal power to electricity again uses steam. Our real-life Calcipher is the fire in the belly of the Earth. Carnot told us that increasing efficiency is a matter of increasing the temperature difference between the heat reservoirs; if you place water under pressure you can increase its boiling point – or even remove it, by converting the water to a supercritical fluid. Supercritical water is already being adopted in a new generation of power plants to increase their efficiency by up to 50%, while in 2014 the Australian government announced the first successful generation of supercritical water from solar power.

Carnot's insight, based on his childhood fascination with a waterwheel, was essential to the development of thermodynamics. The subject was eventually codified in laws, to which we turn now.

You're playing a game

I said there are three-and-a-half laws of thermodynamics: there are three, plus a zeroth law. The number zero suggests two things. First,

---

* Carnot's analogy of a waterwheel is actually slightly misleading, as the Carnot efficiency only applies to engines extracting useful motion from heat.

that this law is of a vital importance which must be established before the other laws can be considered (or it would be numbered four). Second, that this importance is subtle and easy to miss (or it would be numbered one). After the other three laws were in place, scientists realised they needed to formalise a certain assumption they'd been making all along. As Ginsberg stated, we need to establish that we're playing a game before we can establish its rules.

The formal statement of the zeroth law is that:

If system A is in thermal equilibrium with system B, and system B is in thermal equilibrium with system C, then system A is in thermal equilibrium with system C.

To physicists, a 'system' is whatever is being studied, and it is used in distinction to the 'environment' which is not. A barrier is imagined to separate the two. If a system is in thermal equilibrium its energy is unchanging, and it has a uniform temperature throughout. Two systems in thermal equilibrium, if allowed to exchange heat, would do so equally so that no overall transfer takes place.

My friend Professor Stephen Blundell in Oxford states the zeroth law more simply as 'thermometers work'. This seems pretty obvious – thermometers are *boring*: of course they work! But remember, being a wizard is about maintaining that pre-familiarity fascination. Thermometers measure temperature. The scientific unit of temperature is the kelvin (K); this has the same increments as centigrade, but on the Kelvin scale zero is set to 'absolute zero', the theoretical coldest possible temperature.

To see how the zeroth law is required for thermometers to work, imagine you are the type of wizard who goes fishing. You keep a thermometer outside your window, so you can tell at a glance if it is freezing (in which case fishing's off). The operation of the thermometer relies on the fact that the thermometer is at the same temperature as the air; but knowledge of the air's temperature is only useful if it is the same temperature as that of the surface of the water in the lake. So you rely on the fact that the thermometer is in thermal equilibrium with the air, and the air is in thermal equilibrium with the surface of the water. But you also implicitly require that the thermometer would therefore be in thermal equilibrium with the water if they were brought into contact. That last assumption is the zeroth law of thermodynamics: thermometers work!

To see how things might have been otherwise, imagine trying to organise a zoo on similar principles. You put a tortoise in with the lions and see that nothing much changes – equilibrium is maintained. You put the tortoise in with the goats and equilibrium is maintained. So you deduce that since the lions are in equilibrium with the tortoise, and the tortoise is in equilibrium with the goats, it's fine to put the goats in with the lions – it does not go well. A more familiar example is stone, paper, scissors: paper beats stone, and stone beats scissors, so surely paper must also beat scissors. But actually scissors beat paper. While winning an individual game of stone, paper, scissors doesn't have much to do with equilibrium, you might imagine a group of people playing many rounds of the game; assuming there are no psychics in the group, the overall scores would come out about even because each object beats, and is beaten by, one other. If you want to put this to a nefarious use in your local tavern (where by now you will be making a name for yourself) you can manufacture yourself some 'intransitive dice', shown below.

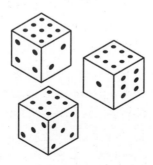

15: Intransitive dice.

Each number is repeated on the opposite face. By working out the possible outcomes of two-dice rolls you can find that each die has a 5/9 probability of beating one of the others, and a 5/9 probability of losing to one of the others. So if you ask your drinking companion to roll first you can always pick a die which is likely to beat theirs. This is bizarre – you'd expect that if A beats B on average and B beats C on average, then A must beat C on average. But actually it loses.

In light of the existence of such tricks, it was not necessarily obvious that thermometers should work. And yet they do, and, as a result, temperature is a meaningful concept. That's good, because hot and cold are pretty intuitive ideas; but it turns out they can also be quite tricky. For example, what is the temperature of a single molecule? To see how counterintuitive temperature can be, please come with me now on a journey through time and space.

Magic carpet ride

As we all know, it's very cold in space. (In case you were wondering whether this is the trick, don't worry, it really is cold: the temperature in space, away from stars and things, is about 3 K, 3°C above absolute zero.) Films have taught us that two things tend to happen when people go into space without a space suit: they explode, and they freeze. The reasoning behind people exploding is that on Earth there is a huge pressure on us from the atmosphere, so the insides of our bodies must balance this with a huge outward pressure. If you remove the atmosphere there is nothing to balance this outward pressure and we burst. But in fact skin is strong enough to contain the outward pressure without too much of a problem – just a bit of bruising. Provided the pressure change is slow enough we wouldn't explode. So let's imagine flying up to space at a leisurely pace on a magic carpet; we can cast a magic spell so we can breathe and not be pummelled by space debris. Do we need to cast a spell in order not to freeze? I don't think so, and I'll explain why.

While your temperature is higher than that of space, you are not able to efficiently exchange heat with space. If you were dropped into a vat of liquid near absolute zero, as happens to Scaramanga's henchman at the end of the Bond film *The Man with the Golden Gun*, you would certainly freeze. You would lose heat by several methods. Conduction: you are in physical contact with the liquid so can transfer heat to it directly. Convection: as you heat the liquid around you it becomes less dense and rises, with cooler liquid replacing it to transfer more heat away from you. Radiation: you constantly exchange infrared radiation – light which has too long a wavelength to see – with your environment. If your temperature is higher than that of your surroundings, you give out more heat by radiation than you receive. In hot environments humans mainly lose heat by a fourth mechanism, evaporation, in which water molecules (sweat) leave our bodies, taking energy with them.

Space is not a vat of liquid, and there is nothing to conduct or convect the heat away from you. Your body would be trying to conserve heat, so would be minimising evaporation. You do lose a lot of heat through radiation: I was once wandering across the barren bogs of Dartmoor with a friend, when he developed hypothermia. While I was waiting with him for an ambulance I too became hypothermic. The paramedics wrapped us in paper-thin metallic blankets which reflected our own radiated heat back into us; almost immediately, I felt comfortably warm. A quick calculation suggests you would lose around three megajoules of energy per hour through radiation into space; my cereal box recommends a daily allowance of 8.4 megajoules of energy in the form of food, so you would need to eat around nine times the usual amount of food in order to take in enough energy to stay warm in space. If we pack some sandwiches on our magic carpet we should be OK.

In terms of conduction and convection, space would feel like a lukewarm bath. But you would not be receiving the usual balance of radiation from the environment: in that regard it would feel like 3 K. So temperature is not as intuitive as it might first appear. The issue is that, in space, you are not in thermal equilibrium with your surroundings: you're losing energy and having to top it up with the energy from sandwiches. But without thermal equilibrium, temperature may not be well defined. Most systems at best only ever approximate such equilibrium: if the air temperature drops when planning a fishing trip, the thermometer and lake each take a while to adjust; this is quick for the thermometer, but slow for the lake. In England the air temperatures are hottest in July but the sea is warmest in late August or early September. So thermometers only approximately work, and only in cases where we have approximate thermal equilibrium.

The concepts of heat, temperature, equilibrium and so on refer to things on our everyday macroscopic scales of length and time. But the macroscopic world emerges from the collective behaviour of large numbers of atoms and molecules. The study of how large-scale thermodynamics emerges from the small-scale world is called statistical mechanics. Let's see how it works for the zeroth law.

The emergence of the zeroth law

Steampunk finds its setting in a divided world. There is the clean, well-to-do upper world. Then there is the dirty, seedy underbelly. Often this is a literal division between above and below. Studio Ghibli's *Laputa: Castle in the Sky* introduced many steampunk tropes; it starts out in a coal-mining town, and ends up on a magical utopian flying island. It is the focus on the dirty lower world which is the essence of the steampunk aesthetic. Mirroring the Industrial Revolution, a luxurious and decadent society is made possible because the dirty work is hidden out of sight in the engine room below.

So it is with our own luxurious middle realm: the work happens down below, in the microscopic world from which it emerges.

These days the idea that our world is built from atoms is so familiar as to seem obvious: we have photographs of atoms taken with scanning tunnelling microscopes. Yet the universal acceptance of atoms is surprisingly recent. Einstein's work in 1905 led to experimental confirmation culminating in the 1926 Nobel Prize for Physics being awarded to Jean Baptiste Perrin 'for his work on the discontinuous structure of matter'. While the idea of atoms was ancient, it had fallen out of favour amongst some philosophers and physicists in the late nineteenth century. This was due in large part to the success of thermodynamics: with its smoothly varying temperatures and flows of heat, there seemed no place for a discrete microscopic world. Condensed matter physics was almost over before it began; the person we have to thank for putting us back on track was Ludwig Boltzmann.

Born in Vienna and descended from clockmakers, Boltzmann was renowned for his fastidious attention to detail. For example, deciding his children needed more milk in their diet, he once bought a cow at a farmers' market. Granted, this is not a particularly fastidious action in itself; but whereas a mere mortal might consult a farmer on how to milk a cow, Boltzmann instead consulted a professor of zoology. That demand for rigour characterised his career, and resolved some rather tricky philosophical issues that had arisen at the end of the nineteenth century.* Boltzmann had an idea: perhaps the smooth macroscopic world around us emerges from the small-scale behaviours of individual atoms behaving according to the familiar laws of Newtonian mechanics. While the behaviour of any individual atom would be impossible to predict, the atoms' *collective* behaviour might be expected to conform precisely to statistical predictions. This is due to the law of large numbers: statistical

---

* There seems to be no record of the professor of zoology's advice; probably it was to ask a farmer.

averages approach their expected values as the number of trials becomes very large. For example, any given roll of a die could show any face, and over a few trials you might expect to see one number dominate just by chance. But the longer you keep rolling, the more equally represented each face should become – the law of large numbers is essentially the assertion that this is a fact about probability rather than a fact about dice. A famous historical example (which I learnt about from one of Derren Brown's magic shows) was provided when statistician Francis Galton* looked at a spread of 787 estimates of the weight of edible meat in an ox by attendees at a Devon agricultural show. While individual estimates varied widely, as might be expected from a large crowd including many non-experts, the mean estimate of 1,197 pounds was exactly correct, to the pound.[5] A central concept in condensed matter physics is the 'thermodynamic limit'. This just means that the number of particles under consideration is so large that the law of large numbers applies to them. It is the limit in which our middle realm emerges from microscopic statistical mechanics. Indeed, a common definition of condensed matter physics is that it is the study of phenomena which emerge in the thermodynamic limit.

The simplest model connecting statistical mechanics to thermodynamics is that of the 'ideal gas'. An ideal gas obeys a simple relationship: its temperature is proportional to the product of its pressure and volume. On the microscale this fact can be explained by assuming that the constituent molecules move independently of one another: they just shoot around bouncing off the walls of the container. When a gas molecule hits the inside of a balloon it bounces back, and, in so doing, transfers some momentum to the balloon, just like how, when you bounce a tennis ball on a racquet, the ball bounces and you feel a push into your hand. The huge number of individual collisions between gas molecules and the balloon at any instant leads to the balloon being pressed outwards. This is

---

* Cousin of Charles Darwin.

the microscopic explanation of why the balloon remains inflated. To understand how our familiar world emerges from this kind of churning activity, it is convenient to think in terms of 'macrostates' and 'microstates'.

The macrostate of a system is its observable, measurable properties: volume, temperature, pressure and so on. Each macrostate is generally consistent with a huge number of microstates: the collection of particle positions and velocities on the microscale. The zeroth law tells us about systems in thermal equilibrium; in statistical mechanics, equilibrium is just the macrostate consistent with the greatest number of microstates. Subject to a few widely accepted assumptions, this makes the equilibrium state the most likely to occur. The zeroth law tells us that a system in equilibrium is as spread and mixed as possible. To take a familiar example, the equilibrium state of air in a room has the molecules evenly distributed throughout the room. This is as opposed to, say, all the air crammed into one corner. Similarly, if two systems are allowed to equilibrate, the macrostate they will tend to reach is the one consistent with the most microstates: they are then said to be at the same temperature.

Just as the knot-makers in the islands west of the westernmost point on the map construct their knots, physicists construct their mathematical models. And as with the knot-makers, for a model to achieve that rare accolade of being considered truly beautiful it must combine practicality, necessity, aesthetic appeal and simplicity. Statistical mechanics is a beautiful theory. The connection it developed between the microcosm and the macrocosm became the basis for condensed matter physics.

According to Ginsberg, the zeroth law establishes some ground rules of the game we're all playing. The remaining laws tell us how it's played.

You can't win

The first law of thermodynamics states that:

Energy is conserved, and heat is a type of energy.

There are two ideas here. The first is the *Law of Conservation of Energy*: energy cannot be created or destroyed, only converted from one form to another. That's a common trope in magical fiction: when a wizard does magic it puts things out of balance, which is one reason why they make changes to the world sympathetically (Rule II). In *The Colour of Magic* Terry Pratchett canonises this as its own rule, the *Law of Conservation of Reality*: using magic to achieve a task must require at least the same amount of effort which would be required using non-magical means.

The second idea is that heat is one of the forms energy can take. The idea that energy is conserved is due to Émilie du Châtelet (1706–49), a mathematician and natural philosopher (a job title that would probably these days lead her to be referred to as a theoretical physicist). As a teenager, du Châtelet used her mathematical ability to devise a variety of successful gambling strategies to fund her expensive reading habits. To see how her idea works, take the example of a hypnotist's watch swinging back and forth. I don't know if hypnotists really swing watches back and forth, although I did once go to a conference on hypnotism in which a set of people were each given a pendulum and told to hang it still in front of their eyes, then were read a script which caused some of the pendula to start swinging through subconscious movements.* At the bottom of its swing the pendulum has lots of kinetic energy associated with its movement, while at the top of its swing this energy seems to have vanished. But then we're told that,

---

* Come to think of it, on the same visit to St Andrews in which I saw a floating crystal, the department housed me in a guesthouse in which the proprietor, over breakfast, was purporting to tell fortunes by a similar pendulum-based method. I explained that the pendulum's energy was mechanical rather than mystical, and gave what I thought was a pretty top-notch demonstration of moving the pendulum 'with my mind' (via my hand). She said she could see my hand moving whereas hers, she insisted, was not. The jury of breakfasting American golfers failed to return a verdict.

rather than vanishing, it's turned into 'potential energy'. But isn't this just a logical sleight-of-hand? Couldn't we take any old thing which isn't conserved, like the number of ducks in a pond, and say that when a duck leaves the pond it just converts into a potential duck so that the total number stays constant? I'd say the essential difference is that we can build mathematical models for the different types of energy. When we add up the mathematical expressions for the different contributions (kinetic and potential, in this case) we see that the total does not change. Presented with a pond and some ducks, I couldn't tell you how many potential ducks there are, so it's not a helpful idea. Actually there is a pond down the road from me with an enclosure to keep the ducks in; inside the enclosure the ducks have a little house. In this case there *is* a good concept of potential ducks, because if a duck is in its house it is not in the pond, but could be. And indeed the total number of ducks in the enclosure is conserved. Underlying this answer is the idea that the mathematics captures the essence of the physical situation. This was du Châtelet's breakthrough: she established, through measuring the imprints in clay of falling balls, the correct mathematical description of the kinetic energy. It was only then that she could see what others – including Newton – had missed: the seemingly unrelated phenomena of a mass moving, and the same mass sitting stationary at a height, can be combined into a quantity, energy, which remains constant.

But the simple example of a hypnotist's watch can't be the whole story, because a pendulum will actually stop swinging eventually unless it is provided with energy (by a fortune-teller's hand or otherwise). Where has the energy gone? It's neither kinetic nor potential. This is where the first law comes in. The pendulum stops swinging because friction converts its kinetic energy to thermal energy: heat. Some will heat the surrounding air; some will pass as vibrations into the hypnotist's hand; some will be given out as the quiet rhythmic rustle of the string as it puts you to sleep. So actually the pendulum by itself doesn't conserve energy, but the pendulum plus surrounding environment does. This is what the first law tells us.

You can't win the game, because winning would mean getting something (energy) for nothing. Remember, if you want your magic to work, you're going to need to put in at least as much effort as it would take non-magically. Perhaps that is why certain wizards are always trying to enlist the help of other magical beings. Susanna Clarke's fantastic novel *Jonathan Strange and Mr Norrell* concerns the return of English magic in the eighteenth century; the name of the game is to indenture the servitude of a faerie to do one's bidding. J. B. S. Haldane, a renowned geneticist who pioneered the use of statistics in biology, wrote a series of books about a magician called Mr Leakey. The books recount Mr Leakey's escapades haranguing a menagerie of magical beings into carrying out tasks (both magical and menial) around his London home and on occasional jaunts out on his magic carpet; a miniature dragon warms his teapot.

The emergence of the first law from the microscopic world is quite straightforward: energy is conserved on the microscale, and continues to be conserved on the macroscale. Assuming, as Boltzmann did, that the motions of individual particles are governed by Newtonian mechanics, these motions are inherently energy conserving. The first law also concerns heat, which is more interesting in terms of emergence as it is only defined in the thermodynamic limit. To see in more detail how heat emerges, let us turn to the second law. The zeroth and first laws tell us about systems once they have reached thermal equilibrium; the second tells us how they get there in the first place.

You can't break even

Thor, Norse god of thunder, once challenged the inhabitants of the castle Útgarðr to a wrestling match. Thor had hoped to fight the formidable giants of Jotunheim, but accepting his challenge was instead an old woman called Elli. As they began to grapple he found that he could not best her; the harder he fought the firmer she stood, until eventually Thor fell to his knee and lost – for Thor was not fighting an old woman: he was fighting old age itself.

So tells the thirteenth-century *Prose Edda*. Hearing this as a child I recall thinking it would be easy to fight old age, because it would have dodgy knees and a bad back. But someone explained the point: the one thing Thor can be certain never to beat is his own old age. The reason for this is the second law of thermodynamics:

> *No process is possible whose sole effect is to convert heat into useful motion.*

While the first law told us that heat and work (useful motion) are both forms of energy, the second law tells us that there is an important distinction between them. Whereas it is straightforward to entirely convert useful motion into heat, as did Thor and Elli in their wrestling, it is impossible to entirely convert heat into useful motion. What does this have to do with Elli's supreme power? The asymmetry leads to a gradual transfer of useful work into useless heat. Death and decay are unavoidable, and everything tends towards disorder. In the game of life, not only can you not win, but you can't break even: eventually, you must lose.

This sense of loss is encapsulated in the idea of irreversibility. Almost every law of physics, every equation, works just as well in reverse. Crystal balls can roll down hills building up speed, and they can roll up hills and come to a stop. The hypnotist's watch swings left and right. These processes are reversible: shown a video of them, you couldn't tell whether it was being played forwards or backwards. Other than a handful of subtle quantum processes, every law of physics is reversible – except the second law of thermodynamics: useful motion can be converted entirely to heat, but the reverse cannot happen. It is remarkable that this is the only (non-quantum) rule telling us that heading into the future is different from heading into the past. Remove this one law from reality, and anything depicted in a video playing backwards would be just as likely to occur as anything in the video playing forwards. Yet our existence is entirely irreversible: scars bear testament to our past; eggs break but do not unbreak; our world tends to disorder.

Disorder is quantified precisely in physics by the idea of 'entropy'. Entropy quantifies the fact that, while we see and experience the world on the macroscale, we have almost no knowledge as to what is going on at the microscale. Of all the possible locations of air molecules in the room, for example, all I can really detect myself is that the air is not all shoved over into one corner (since I can breathe). But the number of arrangements of molecules compatible with my observation remains truly astronomical. Recall that the sum total of information about the molecules (their positions and velocities) is called the microstate, while the large-scale information (air pressure and temperature, say) is called the macrostate. A larger entropy indicates there are more microstates compatible with the observed macrostate.

The second law tells us that, with time, all systems tend towards thermal equilibrium. This is the macrostate compatible with the largest number of possible microstates: while there are many ways for the air to be squashed into a corner of the room, there are so many more ways for the air to be spread approximately evenly that we always experience the latter. The former case could occur, but is phenomenally unlikely. The second law is a statistical statement, and one that is purely emergent: it is explainable in terms of other phenomena, but it can't be eliminated from the explanation in favour of those phenomena.

Some people find the second law counterintuitive, arguing, for instance, that the existence of complex, ordered life seems to violate the trend to increasing disorder. Complex stuff such as bats came about from a messy primordial soup. But while a complex living object such as a bat is in a low-entropy state (there are many more ways for a bat's constituent atoms to be a big pile of bat mess than there are for them to be a functioning bat), the process of the bat growing in its mother's womb created a much larger amount of entropy, as she ate more food to use more energy, some of which she gave off as heat. Life is the process of taking in energy and converting it into heat in order to remain out of thermal equilibrium (i.e. death): equilibrium would be a load of bat mess, not a bat. To put it another way, all change takes place in accordance with the second law, which pushes inexorably towards equilibrium: once equilibrium is reached, the second law's work having finished, there is no possibility of change, or life.

The second law therefore put paid to the quest for the philosophers' stone, the secret of immortality; to live forever would be to stay perpetually out of equilibrium, but since energy converts to heat there would eventually be no energy left to convert. With hindsight, this, along with many an alchemist's aim, was bound to fail. But it is not always so clear whether one's aim is legitimate science or forbidden magic. Fortunately there's a simple incantation which, appropriately used, settles the matter in two words.

Vain chimeras

Many a scientific idea has been shot down when someone points out that it violates the laws of thermodynamics. I've certainly had ideas shot down in this manner. Once such a violation is established, the traditional incantation to exorcise it is the two-word Latin phrase 'Perpetuum Mobile!' (perpet-you-um mo-bee-lay): or, if you're feeling less pretentious, 'perpetual motion'. Identifying that a proposal leads to perpetual motion is an easy way to see that the idea must be wrong, because if perpetual motion exists, then we must be wrong about pretty much everything else.

An equivalent phrasing of the first law is:

No device can exist which creates useful motion without using energy.

Such a device would be a perpetual motion machine of the first kind, violating the first law. Its impossibility is a restriction on the types of magic a wizard may perform. For example, I recently had a thought that perhaps you could make a wind-up mechanical rabbit which, in its motion, wound its own spring. Then it could keep going forever! But I quickly realised that such a rabbit would be impossible as it would get something for nothing. *Perpetuum Mobile!*, I chanted, and the idea was cast back into the abyss of overactive imagination. A more plausible variation was suggested by a friend who asked why electric cars don't use their brakes as a kind of dynamo to charge their batteries, removing the need for a power source. Electric cars do regain some of their energy this way, but again they cannot regain it *all* this way or they would get something for nothing.

In 1670 John Wilkins, Bishop of Chester and a founder of the Royal Society, identified three broad themes for designs of perpetual motion machines, all of which sound very wizardly. They are 'Chymical Extractions', 'Magnetical Virtues' and the 'Natural Affection of Gravity'. Even in the seventeenth century he knew they could not work, yet attempts to make perpetual motion machines persist to this day. The Intellectual Property Office of the UK government felt it necessary to issue the following statement:

Processes or articles alleged to operate in a manner which is clearly contrary to well-established physical laws, such as perpetual motion machines, are regarded as not having industrial application.

IPO Manual of Patent Practice (2016), section 4.05

This statement, with which it is hard to disagree, was articulated more poetically in Leonardo da Vinci's notebook of 1494:

Oh ye seekers after perpetual motion, how many vain chimeras have you pursued? Go and take your place with the alchemists.[*]

The second law, too, can be rephrased as a restriction on a wizard's spells. Any purported device whose sole effect is to convert heat into useful motion is called a perpetual motion machine of the second kind, and is impossible. For example, you might imagine some process in which the molecules in the air hit a ratchet-wheel causing it to rotate, with the net effect being a cooling of the air and spontaneous rotation of the wheel. The second law rules such a device to be forbidden magic.

However, one physicist came very close to devising a perpetual motion machine of the second kind. Like Misters Norrell and Leakey, he did so with the aid of a supernatural being. The person in question was James Clerk Maxwell.

Maxwell's demon

Einstein once stated that his achievements were only possible because he was standing on the shoulders of James Clerk Maxwell. Maxwell made profound contributions to all aspects of modern physics, from unifying electricity and magnetism in his eponymous equations (thereby explaining the wave nature of light and predicting radio waves), to explaining the nature of Saturn's rings, to explaining colour vision and inventing the colour photograph. Yet he never wrote about demons. In a letter to his friend the physicist and

---

[*] Note that Leonardo assuredly dismisses alchemy centuries before its mainstream dismissal with the development of thermodynamics.

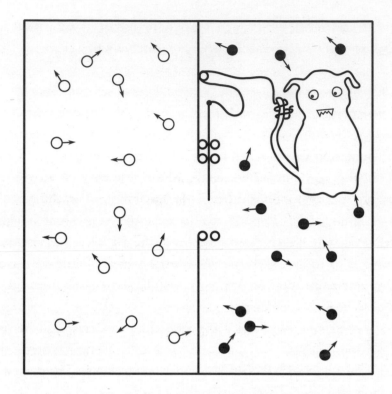

16: Maxwell's demon, as envisaged by my friend Ruth Gordon.

mathematician Peter Guthrie Tait he referred to a 'finite being' with supernatural powers. It was a thought experiment used to push the limits of possibility in the second law of thermodynamics. It was in a letter to *Nature* by Lord Kelvin (of temperature fame) that the demon received its now-immortal name.

Imagine a box divided into two halves (Figure 16). The box can be thought of as isolated, so that it does not interact with the rest of the universe. In one half of the box is a gas made up entirely of black molecules, and in the other half of the box is a gas made up entirely of white molecules. Now imagine a tiny slide window, a few molecules wide, is opened in the wall connecting the two halves.

After a while both halves of the box will contain a grey mix of both black and white molecules. The reason is that there are many more ways for the grey gas to occur than separate black and white gases – there are many more microstates compatible with the grey macrostate than there are with the black-and-white macrostate. If the grey gas were to spontaneously separate into the black and white gases, this would violate the second law.

Now imagine that a tiny demon is able to open and close the window. The demon can also see the exact positions and velocities of all the gas molecules: it knows the microstate of the system as well as the macrostate. Maxwell realised it is possible for such a demon to start from a grey gas in both halves of the box and to separate it into black and white gases, in violation of the second law. All the demon has to do is to open the window without expending any energy when it sees a black molecule passing one way or a white molecule passing in the opposite direction, but not the other way around. Maxwell's original formulation had fast and slow molecules instead of black and white; the demon then separates a lukewarm gas into a hot and a cold gas.

It is natural to wonder whether the demon must do some small amount of work in opening and closing the window, thereby generating heat. That would avoid any issue, as the total heat would still increase. For a long time that was believed to be the solution. But recent experiments confirmed that this heat can be kept much smaller than that lost by the gas; it is safe to imagine the window as opening and closing without energy being expended or entropy created.

After a while the demon will succeed in separating the gases and lowering the entropy; importantly, it appears that there has been no corresponding increase in entropy anywhere else. This last point is key to making a fair comparison: if other things have also changed, they might be hiding the extra entropy. So it looks like the demon has succeeded, achieving perpetual motion and beating the second law.

However, one thing *has* changed. After the sorting process, the demon contains in its brain information concerning all the particles it let through the window; this is information it did not have before. In order to make a fair comparison, this information must be deleted – the demon must forget. But to forget something is an irreversible process: you can't unforget!

You might argue that the demon doesn't need a memory, as it could forget about each particle immediately upon measuring: if you were designing a smoke detector you wouldn't give it a memory. But actually a smoke detector does need a memory – just a really small one. Even if it can only detect a single particle at a time, it still immediately forgets that it detected it, and that's an irreversible process.

The second law quantifies our observation that disorder increases with time. Eggs roll off tables and break; broken eggs don't spontaneously reform, even though this would conserve energy. If Maxwell's demon existed, it would act as a time-reversal device. This is the premise of the film *Tenet*. A similar idea appears in T. H. White's retelling of the Arthur legend, *The Once and Future King*, in which the wizard Merlyn derives his wisdom from experiencing time backwards, allowing him to remember the future. Kurt Vonnegut's novel *Slaughterhouse-Five* has a moving description of a war playing out backwards towards decreasing entropy: planes suck bullets out of the dead returning them to life, and draw up the fire from a burning city and store it in bombs which they suck into their bellies before flying home in reverse to have the bombs dismantled into minerals and buried (Vonnegut had been a prisoner of war in Dresden during the bombing of the city).

The information in the demon's brain points to a resolution with the second law. Since irreversible processes are those in which entropy increases, and the act of forgetting is irreversible, the act of forgetting must correspond to an increase in entropy – a release of heat. The heat released by the demon's brain upon deleting the information matches the heat lost by separating the gases. It must, or the laws of physics would be inconsistent.

Szilard's engine

Steampunk and cyberpunk share a do-it-yourself sense of rebellion. It is a perfect demonstration of thermodynamics' universal relevance that it connects the two; a clear example of this is a steam engine powered purely by digital information. This engine was devised by Hungarian scientist Leo Szilard, who wrote his PhD thesis on Maxwell's demon. Szilard also invented the electron microscope, the particle accelerator and helped clone the first human cell; diagnosed with cancer in 1960, he invented his own cure using cobalt-60 irradiation. Improbably, it succeeded.

This is his fascinating idea. Imagine an airtight shoebox filled with air (Figure 17 on the following page). At either end, on the inside, is a movable internal wall which can be pushed to the centre of the box by poking it with a rod. You can squash all the air into the right half of the box by pushing the left partition wall into the centre. The volume occupied by the air is halved, so the pressure is doubled. Letting go of the rod, the pressure of the air on the partition will push the partition back, causing the rod to extend out. Similarly, you can push the air into the left half and let it expand again. You also have the option to fix the partition in place in the centre of the box.

The box works as both a single bit of computer memory and a simple engine. Say that if the gas is in the left half of the box we'll call that a 0, and if it's in the right half it's 1. With enough shoeboxes you could store all the information of a hard drive: a computer stores the letter 'a', for example, as a string of eight bits: 01100001. So you

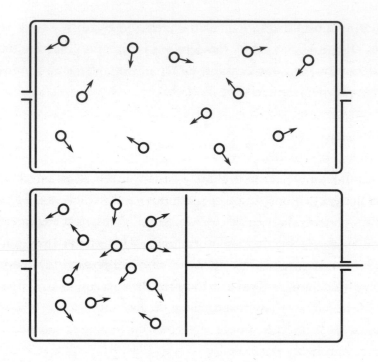

17: Shoebox computer.

could store the letter 'a' in a row of eight shoeboxes by having the gas in the left, right, right, left, left, left, left, right of the boxes, respectively. Any box containing a bit of information can be used to create useful motion, because you can release the partition and the gas will expand and push the rod out. In so doing, though, the information is deleted: after the gas has expanded, the box is neither in state o nor 1. Information is consumed to power the engine.

Szilard imagined a simplified refinement of this. Imagine the gas only contains a single particle. If it bounces around quickly enough it can still create pressure, like a bee buzzing around a paper bag and keeping it puffed out. Imagine Maxwell's demon turns up and can see the location of that particle. It can wait until the particle is in the left half, say, before quickly sliding in the right piston. Since there was no gas in the right half, this requires no energy. The demon can then release the piston, allowing the single-particle gas to expand, pushing the rod out, and doing work. The demon has converted a single bit of information (knowledge of whether the particle is on the left or right) into useful work. You might call it an ... *infernal* combustion engine.

When the single-particle gas expands, its entropy increases. To see this, just note that if the particle is allowed to move around both halves of the box it has twice the number of positions available to it: twice the number of microstates associated with its macrostate. An increase in entropy – disorder – means an increase in heat. In 2010 a team at the University of Tokyo built a real Szilard's engine and showed experimentally that it worked:[6] information can be converted to energy, and deleting information corresponds to a release of heat. This is called Landauer's principle: deleting information, whatever the physical process, creates heat. In the operation of a computer, information is being deleted all the time. With the number of processes running at any instant in modern electronics, the amount of heat being dissipated is huge. Admittedly, there are much larger sources of heat in existing computers, and none yet runs close to the theoretical maximum efficiency set by Landauer's principle. Nevertheless, the laws of thermodynamics tell us that heat production must be a part of any irreversible process. Huge cooling facilities are needed to remove the heat from the servers used for search engines, for example: Google conservatively estimates that a single search on its site uses around 1 kJ, enough energy to power a 35 W residential streetlight for thirty seconds. As of 2021 Bitcoin mining uses so much computing power that, if it were a country, it would be in the top thirty for energy consumption, using more energy than the entire population of Argentina. By design, this usage will continue to increase.

Remarkably, a computer process need not involve deleting

information. The reason I find this remarkable is that even some basic elements of logic are irreversible: the statement 'either I am wearing my wizard robes or I am not' must be true, but the truth of the statement cannot be reversed to deduce whether I'm wearing my robes. Computers enact such logic electronically via 'logic gates' with two wires going in and one coming out: if either in-wire has a high voltage, then the out-wire must, but knowing that the out-wire has a high voltage doesn't allow you to deduce which in-wire does. Yet reversible computing already exists. Quantum computers, for example, operate entirely reversibly except at the final stage of reading out the result of the calculation. Reversible computing sidesteps the Landauer limit, and sits just on the allowed side of the second law. It is practical magic which promises huge gains in efficiency.[7]

As a demonstration of how the laws of thermodynamics unite all of physics, consider what happens if Maxwell's demon tries not to forget. The second law does not require the *recording* of information to generate heat; it might, depending on the particular recording process, but there is no fundamental principle telling us it must in all cases. However, it turns out there *is* a maximum density at which information can exist in the universe. It is achieved only by black holes, and is known as the Bekenstein–Hawking entropy. If the demon attempts to store too much information in its head, its brain must eventually collapse to a black hole. Using the Bekenstein–Hawking formula and the average area of a brain, you can estimate the limit on how much information can be stored before a person's brain collapses into a black hole: I estimate it to be around 10^{70} bits – one with seventy zeroes after it. At the risk of this book seeming outdated in a year or two, I'll say the average storage capacity of a laptop is around one terabyte, meaning a brain could hold about the same information as a little over a billion trillion trillion trillion trillion laptops. Of course, in reality, if there is a reality for demons' brains, the brain would probably just have a finite storage capacity proportional to its volume: once full, the demon wouldn't be able to fit in any more information.

The second law of thermodynamics suggests a rather pessimistic prediction: our world is heading towards a maximum entropy, maximum disorder state, that would support no ordered structures of any kind. This concept is called the heat death of the universe. The idea began to be discussed in the mid-to-late nineteenth century; I recall hearing an interesting theory that it was a natural product of the society in which it developed – western Europe in the build-up to the Great War, when Carnot's vision of efficient engines for spreading empire and destruction was being realised to its fullest extent. These days, it is thought that the fate of the universe will be decided by cosmological factors such as the future expansion of the universe.

Maxwell's is not the only demon with a hand in the development of thermodynamics. A decade before the publication of Carnot's *Reflections on the Motive Power of Fire*, Pierre-Simon Laplace was having rather demonic thoughts of his own.

You can't stop playing

Like Maxwell, Laplace never wrote about demons. In an 1814 essay he imagined an 'intellect' knowing for an instant all the positions and velocities of all the particles in the universe, and observed that this momentary glimpse ought to grant perfect knowledge of all past and future events. The nineteenth century was a good time to be a demon, however, and the intellect soon became known as such. Laplace's demon is often summoned in conversations regarding the emergence of the second law. One line of argument is that entropy effectively banished Laplace's demon from our middle realm: the idea that you can infer the past from the present relies on reversibility, but this is incompatible with our everyday experience (eggs do not unbreak). The debate is still alive; in particular, the following remains one of the biggest unanswered questions in physics: how can the microscopic world be reversible while the macroscopic world that emerges from it is not?

This is called Loschmidt's paradox. There are part-answers: together, the demons have taught us that the second law is only true on average, with many particles, although it only takes a few particles before the chance of the second law being violated is very small. On our everyday scales we never see such violations, in much the same way that you might toss a coin and get two heads in a row but you almost certainly won't get ten heads in a row. Loschmidt was a teacher, colleague and friend of Ludwig Boltzmann; it was Loschmidt's paradox of emergence that led Boltzmann to understand entropy in terms of probabilities.

Laplace's demon cannot exist in our middle realm because entropy is ever present. This last statement is the third law of thermodynamics. One way of phrasing it is:

The entropy of a system tends to a minimum as temperature tends to absolute zero

while another is:

Absolute zero can never be reached in a finite number of operations.

In other words, the lowest-entropy state can never be reached in reality. As Ginsberg had it, you can't stop playing the game.

The third law makes reference to absolute zero, the theoretical lowest temperature of −273.15°C which the kelvin temperature scale takes as zero. Kelvin deduced the existence of absolute zero by thinking through Carnot's analogy of heat engines as waterwheels. He realised that if temperature could decrease indefinitely, eventually engines could become more than 100% efficient – Perpetuum Mobile! There had to be a minimum temperature, and any engine that worked by transferring heat to this minimum temperature would be perfectly efficient. He estimated this temperature by extrapolating from measurements on gases at higher temperatures, using the ideal gas model of independent molecules.

Knowing the value of absolute zero enables you to perform a practical bit of magic: you find yourself needing to set out on a quest of great urgency, but your horse is out to grass, so you hop on your bicycle instead. Your task will take you through the searing heat of the northern wastelands. Recalling the ideal gas law, you know that the pressure in your tyres will increase with temperature, and you don't want them to burst. You also don't want to let air out along the way or they'll be flat by the time you reach your destination. How do you deduce the pressure at one temperature knowing only what it is at a different temperature? Absolute zero makes it straightforward: the ratio of the two pressures is equal to the ratio of the two temperatures measured in kelvin. This only works with absolute temperature scales: the same calculation would not work if you used degrees centigrade or farenheit.

Note that the degrees symbol is not used with kelvin. I originally assumed this was an archaic historical quirk, but my friend Montes explained that it is not. It has to do with absolute versus relative scales. Before it was known that there is a lowest temperature, references to temperatures in physical laws had to be references to *differences* in temperatures. If you refer to a difference it doesn't matter what you pick to be zero.* Once it was established that there was

---

* This also explains the relationship between degrees on a thermometer and degrees arc on a circle: an angle refers to a difference between two things, say the difference in locations of two points on a circle (90° if they are separated by a quarter of the circle).

a lowest possible temperature, it became possible to refer to temperature rather than just differences in temperature, and the degree symbol was dropped. Similarly, the third law tells us it is meaningful to refer to entropy, not just differences in entropy. Loosely, absolute zero is the temperature at which all motion stops. The third law makes this statement precise. To see how, consider yet another phrasing:

The entropy of an infinite perfect crystal tends to zero as the temperature tends to absolute zero.

This is perhaps the easiest to understand, both conceptually and in terms of its practical consequences. If absolute zero could be reached in a defect-free crystal, the entropy would reduce to zero. The crystal would be perfectly ordered, requiring, for example, that its constituent atoms are not moving. This idea of entropy as disrupting an otherwise perfect crystal lattice gives a fresh take on the definition of matter: it is the result of balancing a desire for order against the temptation to chaos.

Free for all

In his short story 'The Imp of the Perverse', Edgar Allan Poe explores the human tendency to purposeless self-destruction. A murderer who has gone on to live a fulfilled life without being suspected of the crime one day finds himself unexpectedly blurting out his guilt to a crowd. Why, he cannot say – was it the whisper of a demon sat on his shoulder? Guilt? Or simply the temptation to irrationality for its own sake?

Crystals are the blueprint for order in the universe. Countless atoms appear in a regular structure, each in an identical environment. Each atom seeks to minimise its energy, like a dropped crystal ball rolling to the bottom of a hill. Since the atoms are all the same they choose identical environments. But if that were the whole story, why does ice melt to water? Why is it not always a crystal?

There is a second process at work. Only at absolute zero would a crystal be perfectly ordered. Whenever temperature is non-zero (which the third law tells us is always the case), the atoms will be vibrating. This is a manifestation of entropy. As the temperature increases the atoms gain energy; they agree to this because they also gain entropy. The trade-off is intuitive: when that dropped crystal ball rolled down the hill, why did it not roll up the other hill and then keep rolling back and forth? Because it feels friction with the ground, and dissipates some of its motion as heat (and sound and vibrations). This process is irreversible, and eventually all its motion is lost.

There is an elegant way to capture matter's competing desires to minimise its energy on the one hand and maximise its entropy on the other. We have Hermann von Helmholtz to thank for it. Helmholtz's path to physics was an unusual one: an army doctor, he spent his free time conducting scientific experiments in the barracks. At the time there was some debate as to whether the emerging universal ideas in physics applied to living beings – do animals and humans simply convert chemical energy to mechanical energy, or is there some immaterial 'vital energy' associated with life? Helmholtz advocated the former view, and conducted a careful demonstration that energy is conserved in the process of muscle movement. This was a major step in the rejection of vitalism, after which physiologists were free to apply physical principles to understand living matter. Helmholtz would achieve a range of further advances across physics, and of relevance here is his work in thermodynamics.

Matter has internal energy: Victorians powered their steampunk contraptions by burning coal and releasing the energy within. But the first law tells us that while some of this energy is free to do useful work, such as moving mechanical castles around the landscape, some of it must take the form of useless heat – the vibrations of atoms and so on. Helmholtz's idea was to subtract this useless part away from the total energy. The resulting energy is free to do useful work: it is known as the Helmholtz free energy.* What is matter? It is the structure which emerges when interacting particles minimise their free energy in accordance with the laws of thermodynamics. At low temperatures matter is best able to minimise its free energy by minimising its internal energy, so order emerges (such as ice or other crystals). At higher temperatures matter is better able to minimise its free energy by maximising its entropy. The balance tips at a phase transition: order is lost, and ice melts to water; at higher temperatures still it boils to steam.

Thanks in no small part to Helmholtz's contributions, thermodynamics was well established by the time of his death in 1894. Its phenomenal success had revolutionised human thinking and cemented the foundations of modern science. The stage was set for another revolution which was to come around the turn of the twentieth century.

The art of the possible

The family tree of physics has two branches: the respectable aristocratic family of natural philosophers – university academics who began to subject their theories to the rigours of experiment – and that family of vagabonds, the alchemists, seeking fortune and glory through arcane knowledge. Thermodynamics played a hand in wedding these families. It was a marriage of necessity proposed and facilitated by the demands of industry. But the vagabond branch

---

* For the sake of completeness, the Helmholtz free energy is defined on the condition that temperature is held constant, in a system able to exchange energy but not matter with its environment.

itself had two lineages: natural magic, in which alchemists sought to understand the world through repeatable tests to establish the veracity of wisdom received from the ancients; and the summoning of demons. There proved no room for the esoteric dark arts in the union: demons were forever banished from science, failing the ordeal of trial by repeatable experiment.

Thermodynamics, captured in its three-and-a-half laws, is the quintessential realisation of those hidden roots that connect our understanding. The laws apply from the microcosm to the macrocosm and to everything between: the creation and behaviour of elementary particles is governed by the conservation of energy; stars are giant engines powered by nuclear fusion; and the developing field of nanotechnology raises thermodynamics to an art, constructing molecular machines to carry out tasks such as drug delivery that it is hoped will one day cure many diseases. Thermodynamics is a clear manifestation of the fundamental difference of our middle realm: entropy, say, simply is not present for individual elementary particles; it is purely emergent, and yet it is the essence of our existence. The difference between past and future is defined by it: time progresses towards increasing disorder; reality as we know it is emergent.

Are the laws truly unbreakable? Well, a particularly sage wizard might answer that there is a way to the world, and that it would be wiser to learn and follow that way rather than to go against it. But wizards are rebellious by nature, and our friend might add, with a whisper behind the back of her hand, that, besides, you must first learn the rules if you intend to break them. With sufficient insight the laws become like friendly sparring partners: in testing their resilience we learn more about the world around us.

We have seen how these laws emerge from the microscopic world of atoms and molecules, as described by statistical mechanics. But the nineteenth-century notion of the microscopic world was overly simplified: it assumed atoms and molecules behaved in familiar ways, shooting around in straight lines. This model is surprisingly effective; but the microscopic world is a much more magical place. As experiments began to grant greater access to it, scientists began to realise that the behaviour of atoms and molecules differs fundamentally from everyday experience. Explaining these experiments required unfamiliar and counterintuitive features to be incorporated into models of reality.

Thermodynamics is a progenitor of condensed matter physics, connecting ancient but often flawed ideas to the establishment of the subject as a science in the modern era. By the turn of the twentieth century the wood had been collected and the kindling laid. In 1905 four sparks were struck from the flint of Albert Einstein's mind; the resulting inferno was the creation of quantum mechanics, and it is to this burning core that we now cast our gaze.

V

Beyond the Fields We Know

Walking home along a moonlit woodland path, Lady Long-Ears stopped and pointed with her finger.

'What is this I point to?' she asked Calabash.

'A hedge,' came the reply. Placing her finger between the leaves, Lady Long-Ears asked again.

'A hedge,' came the reply. Lady Long-Ears circled her finger, indicating that it touched no leaves.

'A hedge includes gaps!' Calabash protested.

Lady Long-Ears explained. 'In these woods there is a name for the tunnels made by birds in a hedge. The word is smeuse.'

Upon hearing this, Calabash no longer saw a hedge with gaps. Instead, he saw smeuses separated by leaves.

❧

The quantum realm

We began with the primordial glimpses of condensed matter physics: the classical states of matter, and humanity's early wonder at lodestones. We then passed through the subject's prehistory, the early appreciation of the powers of metals and of crystals. Then we came to one of the subject's immediate precursors, thermodynamics, which cast away the less useful parts of classical thought – demons and suchlike – bringing testable experimentation to the fore; here we saw how statistical mechanics accounts for the emergence of our world from the microscopic world of atoms and molecules. By the end of the nineteenth century the success of these ideas led to a sense that physics had pretty much explained everything there was to know; details were all that remained.

This view proved rather short-sighted.

Many well-studied phenomena lacked any explanation that made sense with the ideas of the time. There was some missing piece of the puzzle, like finding a carrot, sticks and lumps of coal lying in a field before realising they are all connected by a former snowman. For example, at the turn of the twentieth century, lodestones remained purest magic: there was no explanation of how magnets can exist. To take another example, the ideas at that time suggested that all matter, sufficiently cooled, should freeze into solid crystals. If absolute zero is the temperature at which all movement ceases, what else could happen? Yet with the advent of cryogenic cooling at the dawn of the twentieth century it became apparent that helium could not be frozen. Helium never solidifies at atmospheric pressure – not even at absolute zero. Further, the description of phase transitions in the previous chapter, as a balancing act between energy and disorder, suggests they cannot occur at absolute zero (where disorder ought to be absent). Yet examples of zero-temperature phase transitions are now known. To explain all of these phenomena, and many more, required the abandonment of the classical worldview.

The classical period of physics ended in 1905. Physicists refer to 1905 as the *annus mirabilis*, miraculous year. The miracle was performed by Albert Einstein. Over four short papers he sketched a map of a world that had never before been seen. To borrow from that founding work of modern fantasy, Lord Dunsany's *The King of Elfland's Daughter*, there was a whole world lying undiscovered beyond the fields we know. In the first paper, Einstein invented quantum mechanics. In the second he explained how a simple experiment confirmed the existence of atoms using statistical mechanics. In the third he devised the special theory of relativity. And in the fourth he derived the world's most famous equation, $E = mc^2$. Any one of these papers would have revolutionised the way we see the world. How was he able to write all four at once? The four papers have a hidden connection: step across that border, beyond the fields we know, and they all make the same kind of sense. We'd been floating on the surface of the ocean; Einstein found a way to peer beneath the waves. What he outlined in those papers is modern physics.

In people's minds Einstein is the archetype of eccentric genius and the embodiment of theoretical physics. One of my favourite anecdotes about him was recounted by his nephew: Einstein liked sailing, but would only go when there was no wind, because otherwise he didn't consider it enough of a challenge.* Like all great magic, Einstein's work can be appreciated in three stages. First, it's obviously just magic. Everyone knows that $E = mc^2$, but most people do not set out to learn what it means because they think it to be as impossible as learning a spell. Second, if you study his work at university, you learn that many of the equations attributed to him were worked out by others. For example, the equations describing how time and space distort under relative motion were devised by Hendrik Lorentz. Einstein was entirely open about this, but this earlier work often gets glossed over in popular accounts. But then you get to stage three: Einstein's realisation was as magical as it first seemed – but the magic was not fancy maths.

---

* By all accounts Einstein was a terrible sailor.

His breakthrough was conceptual, and therefore even deeper: it required no secret knowledge. He came up with an interpretation of what Lorentz's equations actually meant and in so doing he simplified our understanding of the world. Science is never done in isolation. Asked in 1950 which scientists he respected the most, Einstein named Lorentz and Marie Skłodowska-Curie. Curie's work on radiation and matter laid a major part of the foundation for Einstein's own results. Similarly, modern physics was not made in a year. All of the subjects Einstein covered in 1905 continue to be developed to this day. As details were understood, the objects of study became more complex: it was decades before quantum mechanics could be applied to many particles at once. The result is called quantum field theory. It was when this happened that condensed matter physics, the study of huge numbers of particles and their quantum behaviours, was born.

The microscopic world governed by quantum mechanics is a world of possibilities and probabilities; could-bes and ought-to-bes. Statements of certainty in our middle realm, like 'my staff is here', become slippery in the quantum realm. Down there, your staff can only have a *probability* of being somewhere. As soon as you find it, it has a probability of being somewhere else. Einstein famously found this quantum lunacy too much to accept. In 1926 he wrote of quantum mechanics that:

> *The theory produces a good deal but hardly brings us closer to the secret of the Old One. I am at all events convinced that* He *does not play dice.*
> Letter from Einstein to Max Born, December 1926

Einstein's objections were well founded: to accept quantum mechanics is to admit that our world is more magical than any fantasy author would dare to write. We will skim the surface of this magic in this chapter, occasionally pressing a glass-bottomed bucket beneath the waves to glimpse the depths.

Before setting off on mind-bending adventures it is important to ground yourself. Clench a talisman; when you fall into infinity you can grasp it, recalling its familiar feel and comfortingly heavy mass, and be reminded that you were sane once and will be so again. That talisman will be a simple fact:

> *Quantum mechanics wasn't invented to make the world more magical.*
> *The world is magical.*
> *Quantum mechanics is its simplest possible description.*

The Old One *does* play dice, and we can see this experimentally.

There are many competing philosophical interpretations of quantum mechanics: proposals as to what the mathematical results actually say about reality. These interpretations cannot be distinguished by experiment. I will try to avoid letting a current of speculation sweep us into that murky ocean by sticking where I can to experimental observations; it is not my purpose in this book to answer questions that have eluded a century of thought. Having a coherent philosophy is important, but most physicists who use quantum mechanics day to day just get on with it without worrying about *why* it works, in much the same way that car mechanics don't spend their days fretting over how cars can move in the light of ancient Greek philosophers' arguments that all movement is an illusion. The approach was summarised by renowned quantum mechanic Professor David Mermin at Cornell as 'shut up and calculate'.

This is also *why* we should care about quantum mechanics: because it gets the job done. Remember the talisman – we didn't invent magic for magic's sake. Quantum mechanics is the most precisely tested theory in history; for example, its calculation of the fine structure constant has been confirmed experimentally to 81 parts per trillion. That is far more precise than the classical mechanics used to build houses, or the fluid dynamics we trust when flying in an aircraft. Quantum mechanics is how we know the structure of stars, how we built the lasers used for fibre-optic internet communications, and how we carry out medical imaging. The process of stepping through the looking-glass with X-ray diffraction is pure quantum mechanics. It governs the world of atoms, and is key to unlocking renewable energy sources such as solar power and nuclear fusion. It governs every electronic device in existence: telephones, computers, everything. And it does all this because condensed matter physics is applied quantum mechanics.

In the previous chapters we saw how our middle realm emerges from the microscopic world. But trawling the depths of the micros-cale we reach the quantum realm – and while seemingly removed from the world of our experience, we must understand its mysteri-ous workings if we are to explain many everyday phenomena, such as magnetism. This chapter asks how it is that our classical world is able to emerge from a quantum one.

The life of psi

The word 'quantum' means discrete, separate. Quantum mechanics is a description of the world in terms of discrete particles. I always assumed it must mean something to do with probability; but in fact the two concepts, discreteness and connection to probability, are related.

Consider a crystal of calcite. Calcite splits a beam of light into two. That's fine if light's just a beam, but quantum mechanics seeks to describe things on the smallest scales: what is the beam itself made of? Similarly, Einstein taught us to think of light as made from smallest parts: elementary particles called photons. But this immedi-ately leads us into trouble. If the light beam is a stream of photons, what decides which way each individual photon goes in calcite? Each photon must have a probability of going each way. OK, I hear you say, maybe it's just like the water molecules in a branching river. But actually it's weirder than that.

Between 1908 and 1913 the Geiger–Marsden experiments (also known as the Rutherford experiments) revealed that atoms have small positive nuclei surrounded by negative charge. The experiments involved firing positively charged α particles at gold foil. Very rarely, about one in 10,000 times, the particles would bounce back. Hence, there must be a small positively charged region in the atom, while the atoms overall were known to be charge neutral. It was tempting to imagine the atom like a positive planet orbited by negative moons, but there's a problem with this: an electric charge moving in a circle radiates energy, so if electrons literally orbited like moons they would lose energy and fall into the nucleus, and atoms would fall apart in no time at all. But nor can the electrons be still, because negative electrons are attracted to positive nuclei, and they would fall into the nucleus like an apple falls from a tree. So what do they do?

Quantum mechanics' answer is that the electrons flow around their orbits like the water flows along a river – but the current is a current of probability, a flow of where the electron *could* be. If you looked to see the electron you would find it in one place – but that's because you looked. Look again: in all likelihood the electron will not be where you would have expected if it carried on flowing along an orbit.

This strangeness is captured in the idea of 'quantum superposition'. Consider a familiar object: a coin (perhaps a thick-hewn groat) slapped to the table in a tavern wager. The coin is one thing or another – heads or tails. By contrast, quantum coins can exist in quantum superpositions of heads and tails, meaning they have the possibility of being either when inspected. Superposition has a reputation as incomprehensible magic – but it is actually familiar as a property of waves. In this context, superposition is simply the observation that when two waves meet (on the ocean, say) the result is also a wave. One consequence is that if two water waves pass through one another they come out on the other side unaffected. Another consequence is that any complicated pattern of ripples is just a combination of simple waves. In fact we met this idea before, in the form of Fourier transforms: arbitrary sounds can be recreated using the right combination of pure tones. So superposition is familiar, and it is in this manner that the quantum coin can exist in a combination of heads and tails: it is like a more complicated wave made of two simple waves (one representing heads, one tails).

What *is* magical about quantum superposition, though, is that when you look at the groat it is *either* heads *or* tails. What was obvious in the middle realm became magical in the quantum – because when you look at a complicated wave on the ocean it continues to look like a complicated wave. It stays in its superposition: nothing forces it to appear as one of the simple waves from which it is built. The magic is all in the 'quantum', meaning discrete: the groat is one or the other; when you look for the particle you find it in one place. All the magic in quantum mechanics boils down to essentially two things, and this is one of them – in the quantum realm, measurements give discrete outcomes.

It's important to note that there is a difference between how we assign probabilities to quantum and classical coin tosses. The traditional way to toss a coin in a tavern wager is to flick it nice and high so everyone can see it turning, catch it in your palm and triumphantly slap it onto the table, before slowly peeking to announce the outcome, then revealing it for all to see. During the stage when the coin is slapped, and no one has seen it, there is a 50:50 chance of it being heads or tails. But this probability merely quantifies the onlookers' ignorance of the coin's value: the coin itself is already one or the other. This is the key difference with the quantum case, because a quantum coin is neither heads nor tails before it is measured – it is a quantum superposition of the two, a measurably different scenario.

To get an idea of the difference, consider the following quantum coin toss. Recall the polarisation of light: thinking of light as a wave, you can imagine it like a rope tied to a post with its free end being waggled up and down. The rope is behaving like the electric field in the light. A polarisation filter is then a bit like trying to send the rope wave through some railings: if the rope waves in the same direction as the railings it can pass through, but if it tries to wave at right angles to the rails it can't. Shining unpolarised light at a polarisation filter is a bit like performing a quantum coin toss: the filter will only let half the light through, while the half that had polarisation at right angles was stopped. It's as if each photon is either heads or tails, and only heads are allowed through.

OK, so now take two polarisation filters, one on top of the other (let's call them A and B), and rotate B through a quarter of a turn. Thinking of them as railings, one set is now at right angles to the other. As a result, no light can make it through – because any light that can make it through A can't make it through B. This is an easy experiment to do yourself if you have two pairs of polarised sunglasses: hold them at right angles and no light will get through. In terms of heads and tails, this fits with our classical intuition: A only lets through heads, and B only lets through tails. For the coin to pass A it must be heads, but then it's not allowed through B.

So now here's the weird bit. Take a third polarisation filter, C, and put it between A and B, at 45°. Since no light was getting through AB, it's obvious that none can get through when you add a third filter, ACB. But actually, light *does* now make it through all three! If you don't have three pairs, you can instead use the fact that your phone or laptop screen is polarised, which acts like the first filter. Hold one pair of glasses in front of the screen, and rotate them until no light comes through from the screen. Now put the second pair of glasses between the screen and

the first pair, and you'll see light through them. It's truly astounding.*
If you try to think this through in terms of the quantum coin being
either heads or tails before you look, it doesn't make any sense. With
two filters you know that all the light between A and B is heads, which
is why none makes it through B. But by putting filter C in that region,
somehow some of the light reaching B is now tails after all. It can't have
had a well-defined value, because that value would certainly be heads to
get through A. So quantum probabilities don't just quantify humans'
ignorance of the quantum realm – they do something more.

Probability is encapsulated in quantum mechanics by the
'wavefunction'. This, too, has a familiar classical precedent.
Wavefunctions describe (for example) water waves on the ocean,
telling you everything there is to know about the wave: the height of
the water at each place, and how far along it is in its cycle of going
up and down. The quantum wavefunction serves the same role. It is
traditionally given the symbol ψ, the Greek letter psi.

I have always been fascinated by the magic of quantum mechanics. When I was young I read all the popular science books I could
find about it; I saw the symbol ψ, and I thought it was beautiful.
These books told me that the symbol appeared in the defining equation of quantum mechanics, the maths which the prose was attempting to approximate: the Schrödinger equation. But those books did
not tell me the equation itself; so, in the days before Wikipedia I had
to imagine those symbols.

When I was fourteen I was on holiday in Alpbach, a small mountain village in Austria. I had with me a popular science book on
quantum mechanics: John Gribbin's *In Search of Schrödinger's Cat*.

---

* Unfortunately 3D glasses from the cinema will not work because these filter
for circularly polarised light, whose direction of polarisation is rotating as the
wave travels. This is so that you can turn your head on its side in the cinema
without messing up the picture.

Walking around the town's graveyard, as teenagers are wont to do, I came across a curious grave. It was no grander than any other, but it was well maintained, with flowers and lit candles. And to my surprise the tombstone bore an inscription in the language of mathematics, the language of the making. Here it is.

$$i\hbar\dot\psi = H\psi$$

18: The inscription on Schrödinger's grave.

It was the Schrödinger equation, the mystic symbols those books spoke of only in words. Through pure coincidence I had found myself standing at the grave of its creator, Erwin Schrödinger. I did not know enough maths to have the faintest idea of its meaning, but I cast it to memory like ancient runes. I knew ψ was magic, and this sequence of marks carved in rock was all there was to say about it; more than words could say.

So I have shown you the symbols I saw that day, as they appear on Schrödinger's grave. I will not explain those symbols, although I will try to convey some of their magic to you in words. The focus here, however, will be on how they tell us about the emergent world around us, and why it is seemingly lacking in quantum magic. To return to that renowned quantum mechanic, David Mermin:

> *I always write about physics to make myself understand better, because equations, in a sense, free one from the burden of thinking about the subject.*

The sentiment is particularly fitting here because even physicists do not know the meaning of the quantum wavefunction ψ: they use it in their calculations to alleviate the burden of thinking about what happens in the quantum realm. What is known is that ψ can be used to deduce, through a simple mathematical operation, the probability of finding a particle in a given place. Yet from uncertainties it conjures statements of certainty; as an example, here is a piece of practical magic it can be used for: quantum tunnelling.

Light at the end of the tunnel

Toss your crystal ball at a wall. Look for it on the other side, and you will not find it – for it has bounced back. But do the same in the quantum realm and you may find the orb on the other side of the wall – and the wall perfectly intact. This is the phenomenon of quantum tunnelling.

Many electronic devices exploit the ability of electrons to tunnel through thin barriers. A device called a tunnel diode uses tunnelling to yield a negative resistance to electrical current: increase the voltage and less current flows, contrary to expectation. Computers are built from transistors which push electrons over energetic barriers; recent experimental work has made transistors a hundred times more efficient by having the electrons instead tunnel through the barriers.[8] Tunnelling is the basis for the radioactive decay of atoms discovered by Marie and Pierre Curie: a radium atom decays when protons and neutrons tunnel out of the nucleus. The reverse process, in which the particles tunnel back into the nucleus, is the basis of nuclear fusion. It is how the sun shines, and mastering it on Earth promises a clean, renewable source of energy. Plants may exploit quantum tunnelling to carry out photosynthesis;[9] understanding this in more detail could lead to a host of green, energy-efficient, processes. Condensed matter physicists routinely use quantum tunnelling to investigate the microscopic world. Remember the picture of individual atoms on page 30? Madhavan's group produced this using a scanning tunnelling microscope. The way this works is that a tiny metallic tip, sometimes itself only a single atom wide at the end, is brought close to the surface of a material. A voltage is applied to encourage electrons to move from the material into the tip (or vice versa). This voltage gives the electrons energy, but not enough to jump across the gap separating the tip and the material: rather, the electrons tunnel across. The microscope then measures the resulting electrical current: the stronger the current, the closer the tip is to the material. At higher voltages the tip can even be used to pick up atoms, like a magnetic fishing game, to build structures atom-by-atom. This process lies at the extreme end of nanotechnology.

Now, you might ask why it's called tunnelling. Isn't it just like jumping across a gap? Well, no, it's weirder than that. If you jump across a gap you start and end high up, and just have to not fall down in between (like how Arthur Dent learns to fly in *The Hitchhiker's Guide to the Galaxy*: he gets distracted while falling and forgets to hit the ground). But if you *tunnel* across a gap you start and end low down, and have to cross a high bit in between – requiring energy you don't have. Imagine an initially stationary ball given a small nudge (Figure 19 on the following page): if there is a (frictionless) valley ahead of it the ball can roll down one side, gaining speed which takes it back up the other side. But tunnelling is instead like crossing a hill in between: the ball can't get going. So whatever quantum thing happens, it somehow has a similar effect to digging a tunnel, which would allow you to cross the hill without going up in the process.

One thing I found to be really magical when I learnt about scanning tunnelling microscopes is that they don't just see what's there: they have the ability to see what *could be there*. The electrons in a material have a range of energies; turning up the voltage of the microscope lets you look at what the electrons would do if they had more energy. After all, it detects the electrons outside the material, where they couldn't be unless the microscope were there to detect

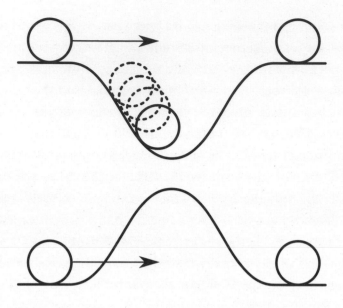

19: Given a nudge an initially stationary ball can roll across a
frictionless valley (top), but not across a hill (bottom). A quantum
ball can tunnel through the hill.

them. I always thought it sounded a bit like that John Carpenter
film *They Live*, where a man finds a pair of sunglasses which show
him an invisible world coexisting alongside our own. But with fewer
zombie aliens.

You can measure quantum tunnelling yourself without expensive
apparatus. One of the best places to see it is your familiar local tavern,
provided they'll have you back at this point. Press a slightly damp
finger to the outside of your full beer glass. View your fingerprint
through the beer: you'll see that it's very clearly outlined, with the
ridges dark and the troughs light. Why is this? Well, the light you see is
reflecting off the inside of the glass. But the ridges of your fingerprints
get so close to the outside of the glass that the photons inside the glass
are able to tunnel into your skin. They are therefore not reflected, and
appear as dark lines. But the effect is very sensitive, and the troughs of

your fingerprints are already too far from the glass for any appreciable tunnelling; hence the light is reflected from these parts.

What actually happens when a particle tunnels? It started on one side of a barrier, was found on the other, and never had enough energy to exist in the region in between. The name given to these types of mysterious processes is 'quantum fluctuations'. The name is misleading because it suggests the particle is gaining and losing energy constantly and randomly, which isn't true: energy is conserved in quantum mechanics just as it is classically. What is happening here is more magical than that.

Tunnelling is one of the clearest examples of the quantum realm's basis in probabilities. To be clear, the phenomenon can also occur in classical waves: there is a classical description of the beer-glass effect, which is that the light takes the form of an 'evanescent wave' connecting the glass and your finger. The quantum mystery comes in only when individual particles are able to tunnel (photons in this case), which has no classical analogue: quantum means discrete. So far the discussion has concerned such individual particles; but one of the most remarkable developments in quantum mechanics since Einstein's work was the realisation that not only can it account for many particles at once, but it *must* do so if it is to remain consistent with one of his other great breakthroughs of 1905, special relativity. The merger of these two ideas is called quantum field theory, and it is the basis of modern physics.

The fields that we know

The 2004 film *I ♡ Huckabees* contains a scene in which Bernard, an 'existentialist detective', outlines his philosophy of reality using a blanket. The blanket, he explains, represents everything: this part, he says, is himself; this is his wife and colleague Vivien; this is a hammer; this, the Eiffel Tower; this, a war. Everything, he claims, is the same, even if it's different. Even in the context of the film Bernard's view is shown not to be the whole truth. But nevertheless, lifting parts of a blanket and declaring them to be different but ultimately part of the same blanket captures the essence of modern physics as depicted by quantum field theory.

To my mind, the easiest way to get an idea of quantum field theory is to return to the idea of phonons. Recall that these are the quantum description of sound travelling through a crystal. A simple classical analogy has atoms represented by little balls, each connected to its nearest neighbours by springs (Figure 20). To simplify, let's imagine a two-dimensional crystal, a sheet of material a single atom thick – like the thinnest possible blanket, or the elastic surface of a trampoline. The reason balls and springs are a good model is that each atom is attracted to its neighbours by their chemical bonds, but doesn't want to get *too* close, because then they begin to feel the repulsion of their positive nuclei. Now, the atoms are not fixed; they can wobble as the springs stretch and shrink. Tug one side of the crystal-sheet and vibrations spread throughout. Lifting one ball pulls up its neighbours by the springs connecting them, and those neighbours pull their neighbours, and so on. If you let go, the balls start wobbling, and vibrations spread out across the trampoline. Now, if the balls and springs are small enough (and remember they're approximating atoms and their atomic bonds), and you blur your eyes a bit, you just see a continuous surface, with a smoothly varying height. This is a field: a description of a physical process which involves assigning a quantity (height in this case) at every point in space. The height of the trampoline at an instant in time and a point in space indicates how much the ball at that point is wobbling back and forth about its resting position.

The name 'field' was proposed by Michael Faraday in 1849 when thinking about magnetic fields (which are another example – the magnetic field describes the direction of magnetic field lines at every point in space). Let me clarify the various metaphors, since 'field' is itself metaphorical. The balls and springs are a classical model of atoms and their atomic bonds. We're thinking of them as a two-dimensional sheet for simplicity (and to enhance the analogy), but the same would work for real three-dimensional materials. Blur your eyes, and they look like the surface of a trampoline. At any given instant a snapshot of the trampoline looks like a set of rolling fields. Chuck a blanket on the trampoline, or a picnic blanket in the field, and it takes the shape of the surface.

The concept of a field was introduced to remove the need for 'action at a distance'. For example, one magnet can influence another at a distance, without touching it. But if this influence were instantaneous it would go against the observation that nothing can travel faster than light – the basis of Einstein's relativity. Fields resolve this issue by saying that magnets both influence, and are influenced by, the magnetic field. Moving one magnet changes the field

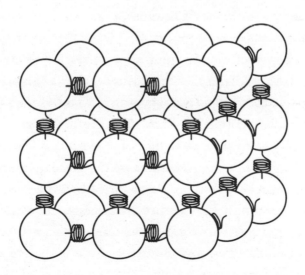

20: Balls connected by springs give a classical analogy for the vibrations of atoms in a material.

at the location of that magnet; the disturbance moves through the field at the speed of light – but no faster – and pushes the blanket up at the other magnet, causing it to move.

To turn a classical field into a quantum field you just have to turn the springs in your model into quantum springs. That turns out to be surprisingly easy; a quantum spring* does what you'd expect: it stretches and contracts like a classical spring. The only difference is that whereas a classical spring can oscillate at any frequency, a quantum spring can only oscillate at certain fixed, discrete frequencies.

Quantum field theory is a description of the world as the collective wobbling of a great blanket woven from quantum springs. One spring exists at every point in space: since we live in three-dimensional space, perhaps a wobbling jelly is a more appropriate image than a blanket. Where the jelly is wobbling quickly there is a lot of vibrational energy: thought of as individual units of vibration – phonons – there are a lot. Other particles such as electrons get their own fields to describe them, but the idea is the same: a concentrated region of energy in the field corresponds to many particles in that place.

To see the power of this approach, consider that all electrons in the universe are identical. They have the same charge and mass, and their only other property, their spin, always has the same magnitude. Why should this be? Why isn't the mass just *nearly* the same? The blanket truth gives a nice explanation. All electrons are identical because they are vibrations of the same quantum field. I have one electron here (lifting up one part of the blanket) and one over here (lifting another). They are the same, in the sense that they are part of the same blanket, but they are different, in the sense that they have different positions. Everything is the same, even if it's different.

---

* The technical name for a quantum spring is a 'quantum harmonic oscillator'.

Quantum field theory is the result of working out the details of the connections between Einstein's 1905 papers. Those papers established that classical physics goes wrong in two different limits: when things get very small, and when they go very fast – to the point where they approach the speed of light. When Schrödinger wrote down his equation in 1925 it described a single quantum particle with mass, such as an electron. This perfectly described what happens on very small scales, but it still assumed those particles to move much slower than light. This is a problem: the Schrödinger equation treats space and time as if they are different things. That's quite natural, of course. But Einstein's work showed that space and time must really be thought of as two aspects of the *same* thing – space-time. These days we have clear evidence of this: for example, an elementary particle which is seen to decay in a fraction of a second when sat still relative to us will be seen to live for thousands of years if we move relative to it. By moving through space we see it live for more time, showing that the two are linked. We see time as progressing more slowly for the particle, and the particle sees time as progressing more slowly for us. While these relativistic effects are fascinating, they are not a focus of this book; the important point is simply that the Schrödinger equation fails to capture them, and so something more is needed.

The first person to write down an equation describing the electron's behaviour that incorporated both quantum mechanics *and* special relativity was Paul Dirac, in 1928.* Dirac explained that when he discovered his eponymous equation he immediately stopped

---

* Dirac was a fascinating character. He turned down a knighthood simply because he disliked being called by his first name – accepting would have required people to refer to him as 'Sir Paul'. He had a reputation for sleeping through seminars, then waking up at the end and asking a question that stumped the speaker. I know at least one eminent physicist who has actively worked to develop the same skill.

working on it and went to bed: he wanted to spend one night believing he'd discovered the relativistic theory of the electron, before waking up and finding the inevitable error. Fortunately for the future of physics, no error was ever found.

Dirac's equation led to some remarkable predictions. First, the existence of another type of particle, identical to the electron in every way except with the opposite electric charge. This is now called the positron, and it is the electron's 'antiparticle'. The positron helped explain a second bizarre consequence of Dirac's work: that a consistent theory of the electron requires the existence of other electrons and positrons – in fact, an infinite number of them! This is certainly a tricky point, but the essence of it is simple enough. Probably the best known of Einstein's 1905 results was the equation $E = mc^2$, which tells us that energy and mass – like space and time – are two aspects of the same thing. If an electron meets a positron they can annihilate, entirely converting their mass into energy. Similarly, concentrate enough energy (say by creating a strong electric field) and you can conjure electron–positron pairs. But here's the thing: measure the electric field around an electron, and you will find the field getting bigger and bigger as you get closer and closer, just as you would if you were to measure the electric field around a charged balloon. But unlike a balloon the electron exists only at a point: it is infinitely small. Measure close enough and you can find such a large electric field, with so much energy, that it allows other electron–positron pairs to be created. This process is called vacuum polarisation (the created pairs are polarised in the sense that they prefer to have their positrons sit closer to the original electron, to balance the charge). So the existence of one electron implies the existence of other electrons and positrons, and the more energy you put into your measurement the more of those particles you will find.

Quantum field theory now underlies our understanding in

almost all branches of physics. Particle physics applies it to the study of elementary particles and their interactions. Cosmologists and astroparticle physicists use it to explain observations on the scale of the entire universe, including the nature of dark matter and dark energy. And condensed matter physicists use it to study the collective behaviour of atoms and molecules within matter, from which our middle realm emerges.

A key result of quantum field theory is that all particles must belong to one of two types, called bosons and fermions. They have fundamentally different behaviours when many of the particles appear together. Phonons are bosons, for example, and some of their behaviours have already been discussed. Electrons, on the other hand, are fermions. These are characterised by their compliance with what is called the Pauli exclusion principle. Now, I'm sure I don't need to remind you of the first rule of time travel according to the 1994 Jean Claude Van Damme classic *Timecop* – you'll no doubt have a well-worn VHS copy, just as I do. It received rave reviews such as 'Van Damme's accent is easier to understand than the plot'. But for the sake of completeness, the rule is that: 'The same matter can't occupy the same space at the same time.' This, in essence, is the Pauli exclusion principle. OK, there are a couple of small inconsistencies with its presentation in that film. For instance, the statement doesn't apply to bosons, which I found to be one of the film's few logical slips. Second, the statement is obviously false: surely the same matter *must* occupy the same space at the same time, by definition. What I'm sure the writers meant was the formal statement of the Pauli exclusion principle: multiple identical fermions cannot exist in the same place at the same time. Remarkably, this simple fact explains why many familiar states of matter are able to exist.

Quantum matters

The late-nineteenth-century description of matter did surprisingly well, but there are certain states of matter of which it simply provided no account. Magnetism is a good example: the Bohr–van Leeuwen theorem (developed by Niels Bohr in 1911 and, independently, by Hendrika van Leeuwen in 1919) is a mathematical proof that magnets cannot possibly exist without quantum mechanics. While the full proof is rather involved, the essential argument is straightforward. A classical account of the behaviour of electrons in a material is given by statistical mechanics. It says that the likelihood of the electrons having any particular mix of motions depends only on their total energy and the temperature of the magnet. But a magnetic field cannot change the energy of electrons. The reason for this is that it causes electrons to move in circles; it only affects their direction of travel and not their speed, so they don't gain energy. Therefore a magnetic field cannot affect the overall collective set of motions of electrons in a material, and cannot, for instance, induce any kind of overall magnetic field. Hence, magnetism cannot exist. Something is missing from the model, and that thing is quantum mechanics.

The Pauli exclusion principle separately gives a straightforward account of another quantum oddity of magnets. The most familiar type of magnet, the ferromagnet, has all its spins pointed in the same direction. But why would neighbouring spins want to align? After all, if you place two magnets side by side, they will try to point in opposite directions.

The explanation is neat, and purely quantum mechanical. Each electron has a spin; the wavefunction describing two electrons must account for their spins as well as the probability of finding the electrons at each location. Now, if the spins point in the same direction, the fermions are identical, and the *Timecop* argument holds: the electrons cannot exist in the same place. But if the spins point in opposite directions the electrons are free to exist at the same location, because they are no longer identical. And if two electrons exist at the same location, there is a large repulsive energy between them, since they are both negatively charged and like charges repel. Therefore, if the spins point in the same direction the total energy is lowered, since they're never in the same place. So quantum spins like to align, contrary to the behaviour of their familiar classical counterparts.

The Pauli exclusion principle is also key to understanding why some materials conduct electricity while others don't. It is remarkable that it took quantum mechanics to provide such an explanation: earlier explanations failed to account for many experimental observations, such as the fact that a material's ability to conduct electricity is proportional to its ability to conduct heat. Pre-quantum predictions all tended to fail in the same way: they predicted that all the electrons in a metal should contribute to its observed properties. After all, if all electrons are identical, what would mark out some of them as special? But experiments seemed to show that only a tiny fraction of the electrons ever seemed to do anything useful. Quantum mechanics explained this discrepancy.

Imagine an argument of wizards (for that, you may recall, is the collective noun) booking into a hotel for a conference. Wizards, like the universe they study, tend to minimise the amount of energy they put into things. And so when the first of them arrive at the hotel-tower hosting the conference they check into the lowest available rooms, to avoid walking up too many stairs. This has the added advantage of allowing them to be first to the breakfast buffet in the tower's lobby, where they will invariably pick all the pineapple chunks out of the fruit bowl. But there's only space for one wizard per room, so late-comers have to start filling higher rooms in the tower. Some of the wizards will be very high indeed. Yet collectively they're all in the lowest rooms they can be; each has expended the least energy possible, but some have expended more than others. Now, it turns out that electrons in materials behave like these wizards: they all want to minimise their energies, but the Pauli exclusion principle dictates that there is only space for one electron at each energy.* And so some of the electrons, like those wizards who checked in late, end up with a lot of energy even in their lowest-energy state. A material is an electrical conductor if it takes very little energy to get an electrical current to flow through it. If it takes a lot of energy to get the electrons to move, the material is instead an electrical insulator. To extend the wizard hotel analogy, whether or not a material conducts electricity is a question that concerns the lowest unoccupied tower rooms. To get the wizards to do something collectively it will be necessary to have some of them change floors. If there are rooms available on the floor immediately above the highest wizards, and the walk's not too far, the material is a conductor, as it takes little energy to get the wizards moving. If, on the other hand, the next available rooms are much higher up in the tower (maybe there is a mezzanine with a stylish corporate art installation), lots of energy is required to get any of the wizards moving. This is an insulator.

And what of this fact that only a fraction of the electrons seem to do anything, such as contribute to electrical currents? Well that's accounted for nicely by the fact that it's only the highest wizards who are free to move. Lower wizards can't move to the floor above them because it's already occupied. Only the electrons with the highest energies in a metal contribute to the electrical conduction or heat conduction. These vital properties of materials could not be understood without quantum mechanics.

This idea need not be esoteric. The red tinge of copper comes about because it absorbs blue photons but reflects red; this is because there is a gap in the possible energies of electrons in copper (a mezzanine in the wizards' tower). Blue photons, being more energetic,

---

* Two if spin is considered.

have enough energy that when they give their energy to an electron it can jump across the gap. In so doing these photons are absorbed. Red photons do not have enough energy and so are reflected.

Another property of matter that could not be explained without quantum mechanics is one of my favourite bits of magic: the Hall effect, discovered by Edwin Hall in 1879. To understand it we must first understand the method by which it was discovered. Take a long, thin sheet of metal and pass an electric current along its length, say by connecting the ends to opposite terminals of a battery. If you were to connect a voltmeter to the same two ends, you would find that it measures whatever voltage is stated on the battery. This makes sense: thinking again of the flow of electric current like the flow of a river, a voltage between two points is like a change in height. The battery acts to create a hill, with a high end and a low end. Just as a river flows down hill, the current will flow from high voltage to low. Now instead connect a voltmeter across the *width* of the sheet. You will find that it records no voltage. Why should it? There is no hill in this direction, as there is no battery connected across the width.

OK, but now pass a magnetic field through the sheet, say by pointing the north pole of a magnet towards the sheet from below. You will find that the width-ways voltmeter now gives a reading. The reason is that the magnetic field has caused the electrons – which previously ran in a straight line along the length – to redirect side-ways. The magnetic field is trying to cause the electrons to move in circles, but the circle is much bigger than the sheet of metal. Which side of the metal gets the higher voltage depends on the charge of the particles carrying the current.

Hall found that the charge of the particles was negative. That makes sense, as electrical currents are carried by negative electrons. Since the charge is always negative, the side of the material that gets the higher voltage (for a given direction of current and magnetic field) must be the same in any material. Right? But here's the thing.

In some materials, the other side gets the higher voltage, as if the current is carried by *positively* charged particles.

These positive particles have approximately the same mass as electrons, but the opposite charge. They cannot be protons, from the nucleus, because protons are much heavier. Positrons might seem to fit the bill, but it cannot be those either, because if a particle meets its antiparticle it will annihilate into pure energy. These mysterious positive particles are present in all sorts of materials, many of which you'll have handled yourself. In fact, basically all metals contain both charges: for example, while indium and aluminium are both metals in the same column of the periodic table, electric currents in indium are carried mainly by positive particles while those in aluminium are carried mainly by negative particles. Other metals with mainly positive particles include lead, tungsten, zinc and about half of the metallic elements on the periodic table. So what are these positive particles? Here's a clue in the form of a classic riddle:

The more you take out of me, the bigger I become. Put me in your pocket and your pocket will be empty. Put me in a barrel and the barrel gets lighter. What am I?

Presence and absence

The key to the mystery is quasiparticles: those particles that emerge through the interactions of huge numbers of elementary particles in real materials – arguably the defining feature of condensed matter physics. Some quasiparticles can be similar to their elementary counterparts; the quasi-electron, for instance, resembles an elementary electron. But others are rather different, with the most extreme cases being those that cannot even *exist* as elementary particles, such as phonons. The particle we seek is of this latter kind. The answer to the riddle above is 'a hole', which is also the name of the particle.

Learning to see holes requires us to learn to see the world as Mr Calabash did upon learning the word smeuse: understanding presence and absence on a more equal footing.* Holes are emergent quasiparticles that only exist in materials. They are very much like positive electrons. The thing I find most magical about them is how they come about. Imagine a line of wizards queueing for the breakfast buffet in the hotel. Each should be wearing their hat of wizardry – it is hotel policy only to serve correctly dressed wizards – but it's early in the morning, and all but one has forgotten their hat. The one at the back of the queue is the only one to remember; they're not going anywhere unless the person at the front has a hat on. So the wizard at the back takes off their hat and places it onto the head of the person in front of them, who does the same, until the hat passes all the way to the front. The hat acts like an electron carrying a current along a wire.† Now imagine it's the afternoon break for the consumption of the elixir of awakening. Everyone's a bit more with it, having already had about fourteen cups of elixir of awakening throughout the day. This time all but one wizard remembered to wear their hat. Typically, though, the one who forgot is first in line, greedily eyeing up the limited supply of pastries. This time the person second in line passes their hat forwards. Third in line passes to second in line, and so on. The end result is that the missing hat passes backwards along the line to the last person.

---

* My friend Jack Winter told me about smeuses. The word is peculiar to Devon, where we grew up. He pointed out that once you know the word, you suddenly see smeuses everywhere, even though you are looking at the same world you saw before. Another friend, Kristen booch McCandless, observed that such tricks control our reality: we see many 'phallic' objects in the world, but we do not see female equivalents, having no word for them in English. The closest equivalent is 'yonnic'; learning the word restores balance, and you see as many yonnic things as phallic.

† Actually each individual electron moves erratically in an electrical current. It is only the average drift of all the electrons which leads to the current.

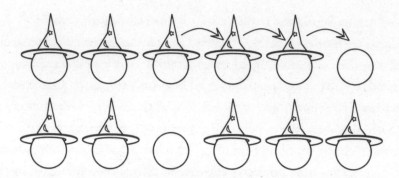

21: Wizards passing their hats forward causes the absence of a
hat to pass backwards.

The situation is pictured in Figure 21. In both cases, the net result is the transfer of one hat from the back of the queue to the front. But in one case a hat moves forwards, while in the other the absence of a hat moves backwards. This second case is how holes work: a hole has a positive charge because it is the absence of a negative electron.

Holes bear some similarities to positrons, the antiparticles of elementary electrons. When a positron meets an electron they annihilate each other. Similarly, dropping an electron into a hole fills the hole, leaving nothing. In fact, this was how Dirac originally envisaged the positron. He imagined the vacuum of space as full of positively charged holes: when an electron is placed in a hole the result is nothing. Imagine pressing a bucket into the ocean to create a bucket-shaped hole in the water. Pouring water into the bucket just fills the hole, leaving nothing. The concept, applied to electrons, is referred to as the 'Dirac sea'. These days elementary particle physicists tend to think of positrons as their own particles, rather than the absence of electrons; but Dirac's interpretation remains valid, and is mainstream in condensed matter physics, where it is referred to instead as the 'Fermi sea'. We've already met the Fermi sea in another guise: it is the hotel tower filled with wizards. Under most familiar situations, when you interact with the ocean you only interact with the water near its surface; most of the deep sea remains unexplored. Similarly, when you interact with the Fermi sea you only interact with the electrons near its surface. Only those wizards in the highest rooms are able to move to higher rooms, and only those electrons at the top of the Fermi sea are able to take part in carrying heat or electricity. When a wizard moves to the room above, they leave an empty room; if another wizard moves into this empty room it becomes full, moving the empty room along.

Holes remind me of a passage in René Daumal's 1952 surrealist classic *Mount Analogue*. In it he tells a tale of Hollow-Men who exist as absences in rock. They feast on nothing, and drink the empty words uttered by their solid counterparts. Daumal informs us that 'as a sword has its scabbard or a foot its imprint, every living man has in the mountain his Hollow-Man, which he will seek out in death'. An electron in a material can similarly seek out its hole-counterpart; filling the hole, the electron returns to the Fermi sea, ceasing to contribute to those properties such as electrical conduction which we measure in our middle realm.

On many a cold and moonlit evening on some far-flung shore you may find huddled physicists keeping warm with tales of adventure on the Fermi sea. I too have warmed myself by such stories. My friend Dr Chris Hooley told me the following tale, or something thereabouts, of the prowess of Dirac in his imaginings:

Once upon a time three friends were fishing on a warm desert island. As always they sought to share their catch evenly. After a hard day's fishing they fell asleep. Early the next morning the first friend awoke. She decided to set off without waking her friends. Counting the haul, she found that it was one greater than a multiple of three. So she returned one fish to the ocean and took a third of what remained. Later, the second friend awoke. They did not realise their friend had left. Counting the haul they found that it was one greater than a multiple of three. So they returned one fish to the ocean and took a third of what remained. Later still, the third friend awoke. He did not realise the others had left. Counting the haul he found that it was one greater than a multiple of three. So he returned one fish to the ocean and took a third of what remained.

Dirac was told this tale, and was asked how many fish the trio had caught.

With patience, trial, and a great deal of error a mere mortal might stumble to the answer: 25. The first friend returns one fish to the sea, leaving 24, then takes a third, leaving 16. The second friend returns one fish to the sea, leaving 15, then takes a third, leaving 10. The third friend returns one fish to the sea, leaving 9, then takes a third, leaving 6. No elegant path to the answer is known. But Dirac is no mortal, for his name is carved indelibly into the great rock of history, just as his equation is carved into the stone of his tomb in Westminster Abbey (Figure 22). And Dirac saw the route to the elegant answer, like the path to the moon glittering on the waves of a halcyon sea. The elegant answer was −2. For then returning a fish to the ocean gives −3, and taking a third leaves −2, and this may be repeated as many times as desired.

22: Inscription on Dirac's grave.

Perhaps a negative fish is a hole in the sea where a fish should go: throw a fish into the hole and the result is a fish-free sea. The story fits a little too well with Dirac's idea of holes in the Fermi sea to take his role in the solution at face value. But what tale worth telling was not embellished with such apocrypha?

Tales of the Fermi sea

Condensed matter physics overlaps with other branches of physics, and with maths, chemistry, engineering and materials science. But there is one thing, at least, that is pure condensed matter: quasi-particles. The Fermi sea provides the ideal setting for understanding them.

The Fermi sea is the equivalent, for electrons in a metal, of the vacuum of space. To use the technical phrase, it is the 'ground state' of the metal, meaning that, while some electrons have a great deal of energy, collectively they have the smallest total energy they can have. Now provide a little more energy, say by applying an electric field. An electron must leave the Fermi sea, moving to a state of higher energy. This is a quasiparticle – an excitation *above* the Fermi sea. Recall the definition of quasiparticles given in Chapter I:

> *An emergent quasiparticle can exist by itself above the ground state of a material, and cannot be reduced to other things with that property.*

One might object that, surely, quasiparticles cannot exist by themselves: doesn't that go against the idea of emergence? But actually that is *exactly* the idea of emergence. While there is a description of quasiparticles – electrons in a material, say – in terms of elementary particles, the quasiparticles cannot be eliminated from the description in favour of those elementary particles without losing something essential. And at the level of description in terms of quasiparticles, these can exist *by themselves* above the Fermi sea.

When an electron jumps, like a flying fish, from the sea to a state of higher energy, it leaves behind the absence of an electron – a hole. The process of conjuring an electron and a hole from the Fermi sea is called pair creation. It requires energy, in much the same way the flying fish requires energy to jump. Pair creation shows that, just like the vacuum of space, the Fermi sea is not really empty: the void is alive with possibility. There is a tricky point here. Every electron in the Fermi sea has the possibility to jump out like a flying fish, leaving behind it the absence of a fish. The fish might be found out of the sea – we might find it on the deck of our ship – but that's only because the deck was there for it to land on. Similarly, the electron might be found with more energy than it could have in the Fermi sea, but that's because the measurement device provided it with the energy. Before measurement the flying fish (electron) and its absent partner (hole) are called virtual pairs. The perpetual probability of them existing is another instance of quantum fluctuation. Virtual pairs account for the possible interactions of the electron with other particles – other electrons, but also phonons, photons, holes and any other quasiparticles inside the material. These interactions lead to measurable effects; one we have seen already is that when an electron enters a material and becomes a quasiparticle, its mass changes.

Quantum fluctuations explain many properties that could not be explained classically. In the last chapter we saw matter as the balancing act between lowering energy on the one hand, and increasing disorder on the other: thermal fluctuations destroy order. By this reasoning, at absolute zero everything should be perfectly ordered: still, solid and crystalline.

This was the view from the late nineteenth century. In 1908 Dutch physicist Kamerlingh Onnes managed to cool helium to such a low temperature that it condensed to a liquid. This was the coldest temperature ever achieved on the surface of the Earth. And yet, no matter how cold he cooled the helium, it never seemed to solidify. How can this happen, when absolute zero is, by definition, completely lacking thermal fluctuations? The answer is quantum mechanical: the material instead experiences the disordering effects of *quantum* fluctuations. To understand this in more detail, let's turn to the most famous act in stage magic.

The cups and balls trick

Houdini is reported to have said that no one could call themselves a magician until they had mastered the cups and balls trick. You know the idea: there are three cups, and under one of them is a ball; you just have to say which – and you're never right. It contains all the essentials of magic: misdirection, sleight of hand, skilled prestidigitation and manipulation of the audience's expectations. It dates back to at least Roman times, but possibly much further; an intriguing mural from 2500 BCE Egypt shows something remarkably similar to a cups and balls trick being performed.

To set the stage, imagine many cups, right way up, each containing a ball. If the cups are all jiggling a bit the balls will be moving inside the cups. Jiggle them a lot and the balls will be flying up in the air, landing in different cups. It's chaos. The amount of jiggling where the balls first start leaving the cups is a bit like a phase transition, separating a low-energy ordered phase (balls in cups) from a high-energy disordered one (balls leaving cups).

Now imagine the cups are not jiggling at all. In fact, make it harder, place the cups face down with one ball under each cup. Surely the balls must now be fixed one per cup. Well, that's the kind of reasoning that loses you the game! Look under a cup: you find two balls. Look under it again: now you find zero, even though there's no jiggling. OK, when a magician does this I admit they use energy. But the universe itself can perform the cups and balls trick, and the only magic it requires is quantum fluctuations. If the balls are quantum particles they could tunnel between cups, effortlessly reproducing the effect which takes a magician a lifetime of practice.

The classical theory of matter has it that phase transitions occur when thermal fluctuations tip the balance between order and disorder. But many phase transitions are now known to occur at absolute zero. These are quantum phase transitions, driven by quantum fluctuations. Remember critical points, those special conditions where matter becomes scale invariant with all its accompanying magic? Well, there are quantum critical points as well, and they are of huge interest to condensed matter physicists, because they seem to accompany many of the most exotic and important phases of matter.

I'd like to be a bit more precise about what I mean by quantum fluctuations. There is a vague sense in which they are the slightly mystifying thing which causes quantum theories not to be classical; they also involve virtual particles whose defining feature is that they cannot be directly observed. But there is also a precise technical sense in which quantum fluctuations appear in theoretical physics. Some physicists prefer the phrase 'quantum corrections', because fluctuations imply something is changing in time, which is not true. The philosophy is this. You find yourself working with a quantum field, as this gives the most accurate model for the experimental observations. But you were raised to believe your world is classical; so you think of your quantum field as classical, but in need of mathematical correction. The classical description of an electron might have it move like a billiard ball; a quantum correction might take the form of the electron throwing out a phonon and catching it again, a virtual interaction between the electron and the vibrations of the crystal lattice. Such processes can't be seen; by definition virtual processes are not measured. They also don't take a finite amount of time to occur, as the corrections are present regardless of how long you leave between measurements. Yet they can affect the electron's properties: they change the classical behaviour so that the electron behaves instead according to the accurate quantum description. Probably the safest way to think of them is by applying Mermin's maxim, 'shut up and calculate': quantum corrections are intermediate steps required in mathematical calculations. If they have a firm physical reality, it is not yet fully understood.

Unmeasurable virtual particles, such as the phonon that is thrown and caught, can have strange behaviours impossible for measured particles. They need not obey Einstein's relation $E = mc^2$; it is in this sense that the analogy between quantum fluctuations and thermal fluctuations works best: the quantum state is thought of as the classical state, but with corrections from virtual particles with different energies. By analogy, matter at a given temperature can have fluctuating regions of higher energy. But the analogy is misleading: thermal fluctuations fluctuate with time, while quantum fluctuations do not. Really the story is backwards: the best description of matter on the smallest scales *isn't* classical, it's quantum. If you embrace that, there's no need to refer to quantum fluctuations, other than as a mathematical step in a calculation. The classical world in which we live then emerges from the quantum.

Quantum mechanics will always be central to condensed matter physics, and quantum field theory will always find a natural home in the study of matter.

Our world, emergent

Modern physics is quantum field theory, the union of Einstein's 1905 ideas. For consistency, even when studying a single particle, it proved necessary to allow for the possibility of many, many more. Some of these are virtual particles that cannot be observed in principle — the moon tugging at the ocean when nobody looks. Others are real intermediates that decay too quickly to be seen. When elementary particles such as the Higgs boson are discovered, they are not measured directly; rather, they decay into other particles that are actually measured. The same is true in condensed matter physics: many emergent quasiparticles cannot exist outside the materials they inhabit; yet the measurement devices, outside the materials, still measure the influence of quasiparticles.

Both elementary particles and emergent quasiparticles are described by quantum fields. It is tempting to think of only elementary particles as 'real' and emergent quasiparticles simply as a convenient shorthand for collective behaviours, but in fact the same mathematics describes both. Furthermore, the idea of elementary particles flying around undetected in the vacuum of space is not something which can be studied scientifically: every particle we actually measure must interact with the matter comprising the detector. The microscopic world from which ours emerges can be a counterintuitive place; but was there any reason to think it should have been intuitive?

For all its mysteries, there are only really two quantum phenomena with no classical precedent. The first is that whenever you look for a particle you find it in one place; the discrete outcomes of measurements is what granted quantum its name. While familiar in our middle realm, this discrete nature becomes magical in a world of quantum superpositions. The second phenomenon is quantum entanglement, and is the subject of Chapter VII.

The first spark of quantum mechanics in 1905 lit a fire whose light ultimately guided us to our current understanding of matter. At this point in the book we transition from the past to the present and future. As a result we will focus on ongoing research: spells that are still being written.

VI

Spells of Division

Within the frantic movement of the library's bookcases, Veryan remained within the small region of calm as she continued to read the forgotten history of the world.

All knot makers belong to a single clan. They are the ultimate historians, archivists, accountants and record-keepers of the world. While the world's wisdom is otherwise passed down through generations by word of mouth or written records, knot makers have a process of encoding knowledge in entwined strings. While usually referred to with the generic name of 'knots'; the alternative word 'nexus' can be used for clarity. A knot maker engaged in the role of historian is termed a *nexus adept*, which is also the name applied to any knot maker skilled in the use of nexuses. A nexus features a central necklace loop off which many further strings hang radially. These radial strings may have further sub-strings connected to them, each of which will have knots of many different types tied in it.

While the nexus can be used as a numerical record-keeping device, a more advanced use has it encode language. The nexus adept reads the nexus by assigning a sound to each knot-type; running their finger along a string, they are thus able to speak a sentence out loud. The pattern of knots on the string also indicates which sub-strings to read, and in which order; in this way the knots can be thought of as punctuation. The nexus surpasses the written word, however, when sub-strings are allowed to connect two or more main strings, creating a complex web. Whereas any written sentence can be encoded by symbols along a line, the nexus is able to encode structures of infinitely greater complexity. The simplest way to understand such statements might be to think of written sentences which contain not only sentiment, but also instructions to jump back and forth to points on the page. Most people perceive history as linear, reflecting the order of the written or spoken sentences used to record it. The knot makers instead understand history in terms of the web-like structure of the nexus.

By an algorithmic process of manipulating the knots, the nexus might also be used as an abacus for calculation. The same can equally be said of the linguistic encoding: a certain knot on a string might indicate a jump to a sub-string, but a different knot, or sequence of knots, might instead indicate that a sub-string is to be detached from one point and reattached elsewhere, or that a new connection is to be made between two strings. In this way the record in the nexus is not static, but dynamic, evolving in the process of reading. The knot makers tend not to draw the same sharp distinction between past and future that the users of sentences perceive. And this is with good reason, for the dynamic nexus is able to encode computations, predicting future events to arbitrary accuracy. This knowledge leads to a further knotting of past and future in the mind of the knot makers, so that they ultimately perceive little distinction ...

ᢙ

Dividing the indivisible

From the outset, the development of condensed matter physics was entwined with that of the computer industry. Despite computers' huge complexity – simulating worlds, underlying modern scientific advances and governing all aspects of our lives – they are all ultimately based on a humble state of condensed matter: the semiconductor. But this chapter is not about semiconductors' rise to prominence: it is about what will be coming next.

Semiconductors lie between conductors and insulators in terms of their ability to conduct heat and electricity. Their magic takes the form of a spell of division, the creation of electron–hole pairs, and their utility is perhaps best illustrated with a simple tale of practical necessity. In the spring of 1944 there was a temporary lull in the fighting of the Second World War around Anzio in Italy. Neither side could make any progress, and soldiers were left with little to do. They were forbidden from listening to the radio because the signal of the powered receiver could be used by the enemy to locate them. One day, Allied soldiers made a remarkable discovery. Connecting one lead of their headphones to a safety pin and the other to a razor blade created a functioning radio – without a power supply.

Here's how this 'foxhole radio' worked. The razor, a metal, had an oxidised layer – rust – on its surface. This metal oxide was a semiconductor. The weak link to the safety pin was acting as an electronic device now called a 'point contact rectifier'. Radio waves, being electromagnetic radiation, naturally create electrical currents in metal wires such as the cables of headphones. But these are alternating currents, reversing direction many times per second, meaning they could not give an audible sound in the headphones. This is where the rectifier came in: it only allowed electric current to flow one way, like a wave sloshing over a wall. This converted the electricity to a direct current which could be heard in the headphones. The rectifier enacted a simple piece of logic: if the current goes this way, allow it; if it goes the other way, do not. This basic idea – that material properties could be used to enact logic – led to the creation of a now famous semiconductor electronic device: the transistor.

A transistor is a semiconductor with three wires attached. Like the rectifier, it enacts a simple logic. Let's call the wires A, B and C. The transistor lets a current flow from A to C if you apply a voltage from A to B. That's a logical statement: if there's a voltage across here, let a current pass there. That simple rule is the basis of electronic computers. We will see how it works shortly.

The success of the computer industry was foretold from the beginning. The prophecy, known as Moore's law, states that the density of transistors on printed circuit boards will double every two years. Roughly speaking, the power of computers will follow the same trend – exponential growth – and it has been obeyed with near-perfect accuracy for over half a century. There are many things at work in the success of this prophecy and it has become, to a degree, self-fulfilling. But Moore's original article in 1965 is full of statements about the physics of the semiconducting matter enabling the technology, observing, for instance, that growth will not be held back by heat generation since the silicon wafers can carry this away efficiently.

Yet growth cannot continue forever in a world of finite resources. This idea is illustrated in an ancient fable regarding the invention of chess recorded by the thirteenth-century scholar Ibn Khallikan. The story goes that the Indian monarch King Shirham was so impressed with the game that he offered its inventor, Grand Vizier Sissa ben

Dhair, any prize. The inventor asked that a grain of wheat be placed on the first square on the chessboard, then twice the number of grains on each subsequent square. The monarch thought this a paltry request for such a fine invention. But when his ministers attempted the task they found that the number of grains quickly exceeded the entire stock of the nation. Accounts differ as to whether the inventor was made a high-ranking minister, or executed for being a know-it-all.

We are the ministers placing wheat on the chessboard: in 2004 scientists in Karlsruhe created the single-atom transistor.[10] There are at most a few Moore's law doublings remaining, and some estimates suggest deviations have already begun. In an economy based on growth, what will happen as we reach the limits placed on technology by fundamental physics?

Moore's law is a statement about a particular *implementation* of computers. To surpass it we must return to the spell of division cast by semiconductors, and ask whether it can be recast in a new form; in this way we might divide the idea of computers from their implementation in electronics. A fantastical example, but one that may not be so far from our own future, appears in the history of the world being read by Veryan, where computing is enacted using the magic of knots.

All electronic devices enact logic by moving electric charge around within semiconducting matter. Gaze as we might into our tea leaves, we cannot foresee what will replace this technology; but one thing is certain: the advance will come from condensed matter physics. In this chapter we will see one possibility: rather than use the charge of electrons, perhaps we can use their other defining property: their spins. It will require a spell of division that condensed matter physicists are still learning; its name is fractionalisation, and it is one of the most bizarre forms of emergence that can occur.

Like many great tales of adventure, ours will take us across the ices in search of the poles. But rather than the frozen landscape of the arctic, our journey will lead us to new types of matter called spin ices. Before setting out into the future, let us first recount the journey that brought us to the present.

The monarch and the magician

A monarch is nothing without a court magician. The monarch uses their power and wealth to facilitate the ethereal studies of the magician, which in turn enhance their patron's wealth and power. Merlin was magician to King Arthur; King Solomon was said to have a court vizier called Asif ibn Barkhiya who was able to travel great distances in the blink of an eye;* Nostradamus was court astrologer to Catherine de Medici, John Dee to Elizabeth I, and Galileo to the Grand Duke of Tuscany. In the same vein, the computer industry gave its patronage to condensed matter physics. In the days of 'solid-state physics' much of the research into condensed matter concerned semiconductor components for electronics and computers. With the financial support of the computer industry, the field at last had the freedom to work out its own esoteric magic. This kind of abstract work often pays the greatest dividends in the long term. A prime example is Bell Labs in New Jersey, founded using the profits of Alexander Graham Bell's invention of the telephone. As a physicist, Bell knew that the key to achieving practical applications lay in studying interesting problems for their own sake. The work at Bell Labs has so far led to five Turing Awards and nine Nobel Prizes for Physics – including the 1956 prize for the invention of the point contact transistor, which underlies modern electronics. To understand how these work in a little more detail it is necessary to understand the state of matter from which they are constructed.

Semiconductors are remarkably paradoxical crystals to look at and hold. Lying in the division between metals and non-metals on

---

* This is possibly the origin story of the magic carpet.

the periodic table, elemental semiconductors have properties that are between the two, which gives them something of an otherworldly quality. They look like rough rocks, but at the same time they're oddly smooth. It's hard to tell if they're shiny or matt – they're somehow both, and neither. To the touch they're neither cold like metals (whose thermal conduction carries the heat from your hand) nor ambient like non-metals (say, this book). Detailed experiments on them prove no less bizarre. In 1833 Michael Faraday observed that the resistance of silver sulphide decreased as its temperature increased, contrary to the behaviour of all known conductors. Metals made sense: more heat means more disorder, hence more resistance. But what was happening in silver sulphide? Faraday had discovered a key property of semiconductors: they cast a spell of division.

Apply a large enough electric field to the vacuum of space, and – in accordance with Einstein's relation between energy and mass – an electron–positron pair is conjured into existence. The same process can happen in matter as well, resulting in the creation of an electron–hole pair from the Fermi sea. Such a spell of division is easily cast in a semiconductor, where a small amount of heat can provide enough energy to create a pair. Because both electrons and holes can carry electrical current, this explains Faraday's observation: increasing temperature increases the number of electrons and holes, and decreases the resistance. What is called an 'extrinsic' semiconductor does not have mobile electrons and holes until the energy is provided to create them.

The best place to observe a hole in its natural environment is instead an 'intrinsic' semiconductor. In a 'p-type' intrinsic semiconductor mobile holes exist naturally (p stands for positive, the electric charge of a hole). An 'n-type' intrinsic semiconductor instead has mobile electrons (n for negative). In both cases the mobile quasiparticles are there because of impurities: for example, starting from pure elemental silicon, a small number of gallium atoms can be substituted in place of some of the silicon atoms. If you look at the periodic table you will find that silicon is in the fourth column, meaning it has four electrons available for chemical bonding, while gallium is in the third column so has only three. Therefore every gallium atom in silicon acts like the absence of one electron – a hole. Substituting arsenic (fifth column) instead of gallium has the opposite effect, adding one electron per impurity. It was this reliance on impurities, along with the close early association of solid-state physics with the development of semiconductor technology, which led Wolfgang Pauli to dismiss condensed matter physics as 'the physics of dirt'. Electron–hole pairs can also be summoned with light or sound. At the start of this book, we saw Veryan speak into her crystal and create light. Her crystal must have been a semiconductor; both light-emitting diodes and laser diodes are semiconductors with n-type regions next to p-type regions, what is known as an np-junction. It is the same mechanism which appeared in the point-contact rectifier of the foxhole radio, making the flow of electric current easy in one direction but hard in the other. Apply a voltage in the appropriate direction across an LED and you create electron–hole pairs: when the electrons and holes recombine they give out energy in the form of light.

An np-junction is not too hard to understand. The entire thing will be a piece of semiconductor, say silicon (although often it will be a compound rather than a pure element). The left half of the silicon contains impurities in the form of single arsenic atoms substituted in for a tiny fraction of the silicon atoms, an n-type intrinsic semiconductor. The right half instead has germanium atoms, and is p-type. You might imagine that, with a load of negative electrons milling about on the left-hand side, and a load of positive holes on the right-hand side, the electrons and holes might seek each other out and annihilate, just as, in *Mount Analogue*, every man has in those rocks his Hollow-Man whom he seeks out in death. And you would be correct: close to the boundary, the electrons and holes migrate to find each other and annihilate. However, recall that the arsenic and germanium atoms are initially charge neutral: they have different numbers of electrons compared with the silicon atoms, but they also have different numbers of protons to cancel the charge. So when the electrons and holes start moving to annihilate, they leave charged impurities behind. The result is that the n-type region close to the border becomes *positive*, and the p-type region *negative*. This establishes an electric field which opposes the flow of further electrons and holes. The silicon further from the division remains neutral. The charged region close to the border is called the depletion layer.

The migration of the charges reminds me of Hope Mirrlees's wonderful 1926 fantasy *Lud-in-the-Mist*. The land of Dorimare, in most regards similar to our own, differs by sharing a border with the land of Faerie. Most people know not to stray too close to the border; there are strange goings on, as some of the Faerie magic has crept into our world. We must assume that close to the border on the other side, some of our world has crept into Faerie, and proves similarly magical to that world's inhabitants. The depletion layer is like this liminal space between the worlds.

The importance of the np-junction lies in what happens when a voltage is applied to it – and the fact that different things happen when the voltage is applied in the opposite direction. Connect the positive terminal of a battery to the p-side, and the negative terminal to the n-side. The voltage from the battery can counteract some of the built-up charge in the depletion layer. The layer gets thinner: more localised to the boundary. The higher the voltage, the thinner the layer becomes, and the electric field of the depletion layer becomes less of an obstacle to the flow of current, until at a moderately low voltage, current can flow across the entire junction. However, with the voltage reversed, the depletion layer instead *grows*. Current finds it even harder to flow: Faerie spreads out into Dorimare, and the Faerie monarch Duke Aubrey, long banished from the land, sets to his wicked and magical ways among the populace.

A transistor is a development on the same theme. The simplest type is a semiconductor with three layers with impurity types alternating npn. Recall that three legs are attached. Leg A attaches to the left (n), B to the middle (p), and C to the right (n). Apply a voltage between A and B and the barrier to current flow from A to C is lowered. Carefully choosing different impurity concentrations in the two n-type regions leads to the desired effect – a simple logical statement: IF there is a voltage from A to B THEN let a current flow from A to C. This simple principle underlies all of modern computing.

Dividing the idea from the implementation

Computers have not always been based in electronics. One argument has it that the earliest precursors to computers were notched bones used as tally sticks or other mathematical aids. Examples include the Ishango bone (made in around 20,000 BCE and found in the Democratic Republic of the Congo) and the Lebombo bone (dating to 40,000 BCE and found in the mountains between South Africa and Eswatini). The Ishango bone has a sharp quartz crystal stuck on the end, making it reminiscent of a magic wand; one sequence of notches contains batches of nineteen, seventeen, thirteen and eleven, leading to speculation that it might have been a calculational aid based on prime numbers. The Lebombo bone has twenty-nine notches, leading to the suggestion that it related to the number of days in a lunar month. It is impossible to tell with current evidence, and it has also been suggested that the notches simply acted as grips.

The abacus was certainly a calculational aid. Operated by moving beads along rods, abacuses were recorded as early as 2700 BCE, in Sumeria. A skilled abacist can quickly carry out complex calculations such as cube roots, and abacuses are still in use today in many parts of the world; I inherited one from my grandad. More mysteriously, in around 100 BCE the Antikythera mechanism was created in Greece. Believed to have been a mechanical orrery capable of calculating the dates of eclipses and other astronomical events, the device contains an incredibly intricate gear system; there is no evidence of anything approaching this complexity until the development of clockwork nearly 1,500 years later.

This hints at the spell of division we might cast to find a route around Moore's law: we must divide the idea of computers from their implementation in electronics. Semiconductors enact logic by dividing positive charge from negative. Might we find some new state of matter which divides something other than electric charge?

Perhaps a natural place to look is to electricity's closest cousin: magnetism.

Charges and poles

While electricity and magnetism have many similarities, they also have one major difference: while electric charge exists, magnetic charge does not. A magnetic charge would be the north pole of a magnet without the south, or vice versa. An electric current is a flow of electric charge; since there is no magnetic charge, there can be no magnetic currents. That would seem to rule out the possibility of a magnetic version of a semiconductor from the outset. But let's examine the idea more closely.

While most objects around us have an equal amount of positive and negative charge, the power of electricity lies in separating the two. If a passing salesperson offered to sell you a magic rag which could rub the electrons off individual atoms, you would surely consider them either a wizard or a charlatan. But such a rag exists: any woollen jumper is an example. Rubbing it on a balloon causes negatively charged electrons to transfer from the jumper to the balloon.

If this weren't so familiar it would be quite magical. The same spell of division is not possible for magnets. But why not?

The simple answer is that a magnet always has both a north and a south pole. The technical way to say this is that all magnets are 'magnetic dipoles', two poles. If you cut a magnet in half you get two smaller magnets, each with two poles. If you keep splitting the magnet, eventually you will find your way down to a single electron which, being an elementary particle, cannot be split. The electron has its own magnetic field – its spin – which again has both a north and a south pole. Since even elementary particles have two poles, it seems there is no hope for a spell of division to separate them.

But there is something strange about this. An electron has a negative electric charge: it is an 'electric monopole', one pole of electricity. A magnetic charge, if it existed – a north pole without a south or vice versa – would be a 'magnetic monopole'. While there are plenty of elementary particles with electric charge, an elementary particle with magnetic charge has never been seen. The weird thing is that there seems not to be any fundamental reason why this must be. On the contrary: the laws of physics would look a lot neater if magnetic charges were found, while their existence would also answer some important questions about the nature of reality: for example, why electric charge only appears in multiples of the electron's charge. There are many ongoing experiments devoted to the search for elementary magnetic monopoles, but they have so far turned up nothing. Given the absence of magnetically charged elementary particles, we cannot have magnetically charged objects.

It is tempting to ask how magnetism exists at all given that there is no magnetic charge. This just comes down to the fact that while elementary particles never seem to have magnetic charge, they do sometimes have magnetic dipoles: this is what is meant by the particle's spin. While classical analogies for spin always fall short, loosely we can think of it like this: if an electric current flows around a loop of wire it generates a magnetic field. The spin of an atom can be thought of as the flow of the electric current as an electron orbits the nucleus. How a single electron can have a spin is even more puzzling, but you can note that if you charge a balloon with your jumper and set it spinning it will again generate a magnetic field as the charge rotates. One thing these analogies capture is why you might always expect both poles of the magnet. Point your charged balloon at a clock. Now set it spinning in the same direction as the clock's hands. If you measure the magnetic field of the balloon from where you stand, you would find a north pole, as you see the balloon turning clockwise. But if you went round to the other end of the balloon you'd measure a south pole, as the balloon would appear to be rotating anticlockwise from your new perspective. Separating the poles of a magnet would seem to be as impossible as separating clockwise from anticlockwise motion.

But despite all this, there is hope for the magnetic semiconductor. For condensed matter physics is not the study of elementary particles: it is the study of emergence.

Sawing a person in half

Of the knot makers assigned to an island, one will be the knot master. This position conveys not only a great skill at the knotting arts, but also wisdom in governance. The selection of the island's knot master falls to the clan of watchers. That this is the watchers' primary purpose demonstrates the importance of the appointment. The watchers have a series of tests which they apply to members of the knot-making clan at a very young age. Some of these are directly knot-related, such as the untying and reproduction tasks which comprise the knot makers' competitions. Other tasks require objects to be passed through holes and around obstacles in certain sequences, including knitting and stitching patterns on inspection. Some tasks take the form of riddles, such as

This string has a left and a right end.
Show me a string with only a left end.

The purpose is not so much to arrive at a given answer as to enjoy the process of weaving the mental knot which the statement begins. Initially resembling a joke or triviality, the knot master may spot in such statements a tiny thread of sense which, if pulled, will begin a grand unravelling without end. Two knot masters engaged in conversation can have such a way of discussing that an onlooker might reasonably believe them to be talking nonsense. In such cases the observer is then confused as to how the masters appear to be understanding one another and arriving at sensible conclusions. In reality the masters are engaged in entwining two lines of thought ...

This book is not intended as an exposé of magicians' methods. It is not the short-lived 1990s television series *Breaking the Magician's Code: Magic's Biggest Secrets Finally Revealed* presented by *The X-Files'* Mitch Pileggi. I wish it were. Mitch's undisguised disdain for the tricks, the 'Masked Magician' performing them, and the audience at home made the show an engaging Sunday afternoon spectacle. However, just this once I will break the magician's code and finally reveal one of magic's biggest secrets: how to saw a person in half.

Or rather, I will finally reveal how I thought the trick worked, because it turns out I was wrong. I assumed there was a second person hidden in the box: I thought I was seeing one person's head and another's legs. In fact Mitch tells us that whenever we see feet sticking out of the box, they are fakes operated by remote control.

Let's propose our own version of the trick which works the way I imagined. In that case, what seems like a single person divided in two is really two whole people. Perhaps you have seen this version performed by a person pretending to magically divide their thumb from its tip: in reality they have secreted, about their other hand, their second thumb. (Either that or it's a fake thumb operated by remote control.) By having multiple copies of an object we can make it appear that a single copy is split in two. In condensed matter physics we have not two copies of something, but a great many.

Imagine laying many magnets end to end. Each has its north pole painted red, its south pole blue. If you blur your eyes enough, you see purple everywhere. Now, flipping one magnet creates a double blue next to a double red. These double regions are large enough that they maintain their blue or red colour when you blur your eyes; in this way, there is a concentrated region of north next to a concentrated region of south. Now flip the next magnet along. The double red, say, moves along one place. By flipping subsequent magnets, the concentrated regions of north and south can move independently (Figure 23). The poles have been divided!

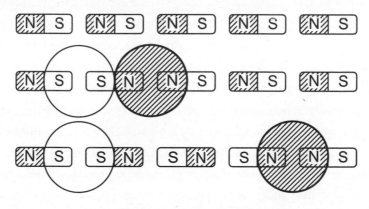

23: Flipping magnets row by row.

To emphasise how impressive this is, here's a simple way to see how impossible separating the ends of a magnet is. Imagine the magnet laid in front of you, its north end to the left. Looked at this way, asking to see the north pole without the south is like asking to see the left end without the right! Separating the poles, even in an emergent sense, has answered the riddle of the knot makers: *show me a string with only a left end*. To isolate the concentrated region of north pole, you'd need a very long string of magnets, and you'd have to push the south pole far away to the right.

Using emergence, we have created magnetic monopoles. A net amount of magnetisation emanates from the double north, and returns to the double south, albeit in such a way that no law of physics is broken. Now, this macroscopic tale of bar magnets is a fun reinterpretation of something that is entirely familiar. But what would be really magic is if the same thing occurred naturally, on the atomic scale, in a real crystal. My reason for saying so is this. First, there is some intuitive aesthetic sense in which things seem better if they're made by nature. Artificial grass is rubbish, for example, and real opals can cost millions of times more than artificial ones. Second, as I hope I've convinced you, emergent quasiparticles are just as real as elementary particles. In this sense, if magnetic monopoles could emerge from the behaviours of individual atoms, they would be real. But what would be most pleasing of all is if monopoles turned out to provide a simple way to understand an otherwise impossibly complicated situation. This is not quite true at the level of flipping bar magnets.

There would also be practical advantages to monopoles coming about naturally. While scientists can manufacture arrays of magnets it takes time and energy, and the results have to be fairly large. But nature can manufacture huge complicated arrays of magnets, on the atomic scale, in the blink of an eye. This is just crystal growth, and the result is far smaller and more robust than the best human-created microelectronics. That would be what is required for a magnetic version of a semiconductor. It is fortunate, then, that nature, with a hint or two from the crystal growers, can perform its own version of the sawing a person in half trick.

A *new spin on ice*

Spells are overheard from the whispers of nature. Where do we find this spell of division? Crystals of dysprosium titanate ($Dy_2Ti_2O_7$) and holmium titanate ($Ho_2Ti_2O_7$) are paramagnets. They are not known to grow naturally, although several crystals of them have been grown by crystal growers around the world. When they are cooled to very, very low temperatures, they turn into something remarkable: 'spin ice'.

This is a fundamentally new type of magnet. Being crystals, the atoms of spin ices sit in a regular periodic structure (strictly they are ions, as they have an electric charge, although this is not essential to the

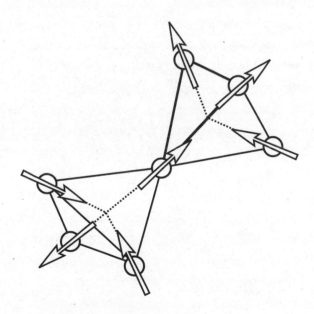

24: Two tetrahedra, triangular-based pyramids, with spins on each corner.

story). Each of them has a magnetic field, a spin. The regular structure of the crystal lattice of the spin ices is built out of triangular-based pyramids: tetrahedra (Figure 24). Each pyramid has four corners. These fit together in such a manner that every corner of a pyramid touches the corner of another. The ions themselves sit on the meeting places of pairs of pyramids. The spin of the ion points directly towards the centre of one pyramid, and directly away from the centre of the other.

Imagine all four spins pointed towards the centre of the tetrahedron on which they live. That is, all four north poles pointing into the centre, towards one another. Magnets wouldn't like to sit like that, because like poles repel. But nor would they all want to point outwards, because then all their south poles are together. The best you can do is to have two point in and two point out. There are six ways to do this per pyramid. Having this two in, two out arrangement on every pyramid in the crystal is called the spin ice state (Figure 25). It is the lowest energy state, the ground state.

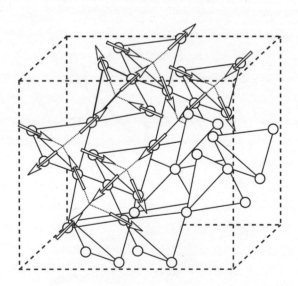

25: The unit cell of spin ice, with two spins in and two out on each tetrahedron.

Now imagine the north ends of the spins are painted red, and the south ends are painted blue. We're on the atomic scale and paint doesn't really exist down here, but you can imagine. If you blur your eyes, two in, two out (two red and two blue) looks purple. So the spin ice state is purple everywhere. But real crystals suffer the disordering effects of temperature, and so not all pyramids find themselves at their lowest energy. Flip an 'out' spin to an 'in', and you now have a three in, one out. With three red ends and one blue end meeting, the result looks red when you blur your eyes. It's a concentrated region of north pole: a north magnetic monopole! But that flipped spin was shared with a neighbouring pyramid, which is now three out, one in. It looks blue – a south monopole. Subsequent spin flips can separate the north and south, just as we hoped.

This ability of the concentrated regions of north and south to separate is the natural realisation of magnetic monopoles we hoped for. Some researchers prefer not to refer to them as quasiparticles on the grounds that their emergence can be explained by the laws of classical physics, rather than quantum mechanics. But such properties do not distract from the reality of the monopoles. For example, if you were to place a magnetometer (detector of magnetic fields) close to the surface of the spin ice, you would expect to see the magnetic field change exactly as it would if an elementary magnetic monopole were there. To reiterate one of the central themes of this book: experiments measure emergent properties, because our reality, our middle realm, is emergent.

The name 'spin ice' has nothing to do with temperature. In fact, the crystals are much colder than ice: they become spin ices below around 2 K, 2°C above absolute zero. That's colder than the universe itself! The temperature of the universe, in the vacuum of space and away from stars and other things, is about 2.7 K. This is the temperature of the cosmic microwave background, residual energy from the Big Bang. It takes work to cool crystals to such low temperatures, but it can be done by skilled experimentalists.

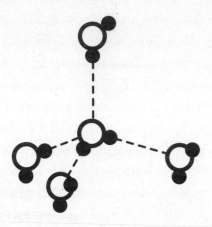

26: The molecular arrangement in ice.

So where, then, does spin ice get its name? Ice is made of water, H_2O. Its crystal structure has the oxygen atoms arranged into a repeating periodic structure. A given oxygen has four nearby oxygen atoms in a tetrahedron around it (Figure 26). Each oxygen has close-in hydrogens along two of these directions. Which two is not fixed, meaning there are again six options per tetrahedron. In terms of hydrogens, it's two in, two out, like the spins in spin ice. This is called the ice rule, and it has some deep implications for the nature of matter.

New worlds beyond the poles

A common theme of science fiction and fantasy has a new world discovered beyond the Earth's poles. Probably the most famous contemporary example is Philip Pullman's *His Dark Materials* trilogy; but it is also the theme of what is often cited as the first work of science fiction, Margaret Cavendish's *The Description of a New World, Called The Blazing-World* written in 1666. A clear inspiration for

Cavendish was the publication, the year before, of Robert Hooke's scientific discovery of a new world of living organisms on the microscale in his *Micrographia*. Spin ices similarly cast their spell of division by opening a door to a new world on the microscale; in this crystal-world there are magnetic monopoles, which seem not to appear in our own.

A key piece of early evidence for the spin ice state was provided by measurements of the crystals' heat capacity. A bump in the heat capacity as the crystals are cooled below around 2 K mean it takes more heat to cause a given change in temperature. This looks a little like what you might see in a phase transition to a new state of matter, but in a phase transition there would either be a dramatic spike in the heat capacity, or a jump, whereas spin ice has a smooth hump. So what causes it?

In Chapter III we saw the textbook definition of matter: that which emerges on the macroscale when the interactions of many particles spontaneously break a symmetry and lead to rigidity. A typical example of this is when a ferromagnet grows from a paramagnet. At high temperatures the spins of individual ions in the paramagnet point in random directions. While they feel one another's magnetic fields, the disordering effects of temperature mean that they prefer to maximise their entropy rather than minimise their energy. As the material cools this balance shifts until, at the phase transition, their spins spontaneously align. The result is termed long-range order. Try to turn one spin and all resist together: the state is rigid.

When a spin ice grows from a paramagnet something else happens. At high temperatures ('high' here in a rather relative sense – above 2 K), the spins in each tetrahedron are pointing in and out randomly. Many tetrahedra have four in, zero out; many have one in, three out, and so on. On cooling the material below 2 K, the tetrahedra all begin finding their way to two in, two out configurations. Given enough time, in principle, all tetrahedra could find their way to two in, two out. But this is *not* long-range order: if I know the orientation of one spin it does not tell me anything about the orientations of spins further away. But there is a *correlation* between spins, in the sense that every spin is part of a pair of two in, two out tetrahedra. And there is something really remarkable about that.

To see why, picture the following simplified model of spin ice (called square ice, shown in Figure 27): draw a grid of squares and try to draw arrows on the edges of the squares so that at every corner of a square – where four arrows meet – two point in and two point out. It's very hard to do unless you have all the arrows doing the same thing (say, every vertical arrow pointing up and every horizontal arrow pointing right). It's quick to try, and you'll see how tricky it is. The problem is that while the first few arrows can be put down randomly, you quickly find that you get stuck, and whatever arrow you draw makes it impossible to do two in, two out somewhere else.

The fact that real materials can solve this problem is impressive. Even more impressive is that they can do so with each spin

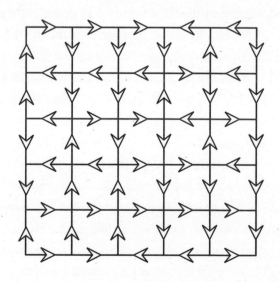

27: Square ice.

caring only about what its immediate neighbours are doing. Like ants arranging themselves into complicated structures simply by interacting with their neighbours, interactions between neighbouring spins lead to correlated behaviour on a large scale. It is a clear example of emergence.*

Incidentally, the spins do still feel the magnetic fields of further-away spins; these just don't change the story. When these interactions are included in the model, it turns out that the emergent magnetic monopoles begin to interact with one another. They do so in precisely the same way that electrons interact via their electric fields, making the analogy even closer.

This idea of long-range correlation without long-range order suggests a more nuanced approach to the states of matter. It is tempting to think of spin ice as a state of matter, as it has measurably different behaviours on the macroscopic scale from the paramagnet from which it grows. But most researchers instead refer to it as a 'correlated paramagnet' and the hump in the heat capacity as a 'crossover' rather than a true phase transition. In the end it depends on how you define all these terms.

What can be said for certain is that the long-range correlations of spin ice have real consequences in our middle realm. Another key piece of evidence for spin ices came from neutron diffraction. This technique is similar to stepping through the looking-glass with X-ray diffraction – switching small to big, to photograph the crystal world. When a liquid freezes into a crystal it develops long-range order; if X-rays were being shone into the matter as this happened,

---

* If you wish to create a two-in, two-out pattern in square ice the simplest way is to use the spell of division: orient arrows randomly, then think of three-in, one-out corners as north poles, three-out, one-in as south poles, four-in, zero-out as double norths, and four-out, zero-in as double souths. Then find lines of arrows pointing from a south to a north: flipping these removes a north and a south, annihilating the monopoles to create the two-in, two-out vacuum.

the diffraction pattern would develop a set of sharp, tiny spots. The size of the spots gets smaller the longer-range the order is in the crystal. Now, neutrons can also undergo diffraction. This is perhaps a little mysterious since neutrons are traditionally thought of as particles rather than waves, but they are really quantum objects and they can diffract. The major difference is that individual neutrons have spins. And so neutron scattering doesn't just probe the crystal structure in a material: it also measures the directions of spins. So what happens to the neutron diffraction pattern as the material cools into the spin ice state? It can't develop sharp spots, because these would indicate long-range order, and there is none. But something small should appear, because it corresponds to a long-range correlation, and large goes to small in diffraction. The features that appear are called 'pinch points'. They stretch in one direction and are squeezed in another. The dark regions in Figure 28 represent intense bits of the neutron scattering pattern; contours show how the

28: Neutron scattering image of spin ice.

intensity changes. Pinch points are the bits where the black regions almost touch (at better resolution, they do).

When I was a child I often wondered whether it is possible to pull the north pole off a magnet. With time I came to believe that it probably could not be done. But the thought persisted in the back of my mind that maybe, somehow, it might be possible. When I was an undergraduate at Oxford I had to choose two options in which to specialise in my fourth year. I had chosen theoretical physics and particle physics. But the summer before I began that final year the news broke that magnetic monopoles had been found in spin ices.[11] The work was theoretical, but involved a clever comparison between numerics and existing experimental data which made a convincing case. Upon hearing this, I immediately changed subjects: from then on I would study theoretical physics and condensed matter physics. When I finished at Oxford I began a second master's degree at the Perimeter Institute for Theoretical Physics. For my research project I used a technique developed in string theory to model neutron diffraction in spin ices. It might not have been exactly what I'd had in mind when I'd wondered about pulling the north pole off a magnet, but it was no less magical. I had returned to that original childhood wonder – it was possible after all, but in a more nuanced way than I had imagined.

The colours of noise

More recently I had the good fortune to be involved, as a theorist, with another experimental search for spin ice correlations, and one which both furthered and clarified the analogy to semiconductors. I would like to tell you about it as the process captures the twists and turns of scientific discoveries. Let me begin the story properly.

We were somewhere around Berkeley, a hundred miles from the edge of the desert, when the science *began to take hold.* My friend Professor Norm Yao had built a device capable of sensitively detecting magnetic fields on the scale of tens of atoms. I suggested using it to look for magnetic monopoles in spin ice. We got in touch with Professor Stephen Blundell in Oxford, my former teacher and an expert on spin ices, and Professor Amir Yacoby in Harvard, an expert on the magnetometry techniques we were hoping to use. The hero of the story was Dr Fran Kirschner, then Steve's DPhil student in Oxford, who used her computer wizardry to conjure some excellent numerical simulations of what we might expect to see.[12] It turned out that at the temperatures which could be reached in the experiment there would actually be *too many* monopoles to detect them. It was as if somebody had theorised the existence of a raindrop, and had predicted the sound it would make when it landed, only for the raindrops to arrive in great numbers and with a *shishhh*.

In fact the listening analogy is quite apt. The magnetometer would detect the strength of the magnetic field at a point close to – but outside – the spin ice. This field would increase and decrease as monopoles moved towards and away from the detector, and popped into and out of existence. With so many monopoles around, the signal would be pure noise. But recall the Fourier transform: any noise can be decomposed into a mix of pure tones, and the same goes for magnetic fields that change in time. So Fran used the Fourier transform to work out which mix of frequencies of magnetic field we could expect. The result really did sound a lot like rain landing, or the song of a river as it flows over pebbles, or the boiling of a kettle: something like *shishhh*. But when she looked into the details she found that it was subtly different: it was a little more like *shooshhh*.

To make this precise it is necessary to understand that not all noise is the same. It has what is called 'colour', in analogy to the colours of light. If you take an equal mix of all frequencies (colours) of visible light, the result is white. And if you take an equal mix of all different frequencies (tones) of sound, the result is white noise. A lesser-known example is pink noise. This mix contains more of the lower frequencies, making it more of a shooshhh than a shishhh. It is again named in analogy to light: if you took a mix of frequencies of light, but with a greater contribution from the lower-frequency (redder) colours, the result would be pink. Whereas white noise is pretty horrible to listen to, pink noise is quite soothing. I first learnt about it while on work experience as a teenager, where I worked for a manufacturer of loudspeaker cabinets. When you make a speaker cabinet, it's important to check it doesn't buzz and you need to test it at all different frequencies. One way to do this would be to play white noise, as that contains an equal mix of all frequencies. But no one would ever want to listen to white noise on their speaker. Instead you can use pink noise, because that has all frequencies represented in approximately the distribution present in music.

One theory has it that people find pink noise soothing as it reminds us of the sounds we heard in the womb, which makes it interesting that the music we create also has that mix. Pink noise appears in a lot of places, in fact. Scientists have claimed to find it in a huge number of natural and unnatural phenomena, from stock markets to tide heights, neuronal firing, DNA sequences, heart beat rhythms and gravitational waves. It is often touted as one of the clearest examples of universality in physics.* If you add even more weight to the low frequencies (like saying *shooshhh* but with a deep voice) you arrive at red noise. Many processes are associated with red noise. For example, Brownian motion, the movement of a pollen grain in water, is an example of perfect red noise. It was by constructing a mathematical model of the pollen grain's motion when buffeted by individual water molecules that Einstein convinced the world of the existence of atoms in 1905. The essential idea had already been

---

* There has also been considerable controversy with the universality of pink noise, with critics pointing out that some analyses are potentially based on flawed statistics.

suggested as early as 60 CE by the Roman philosopher Lucretius, who gave the example of wobbling dust viewed when 'sunbeams are admitted into a building and shed light on its shadowy places'. But it was the precisely testable predictions of Einstein's model that allowed the microscopic theory to be confirmed beyond doubt.

Let me explain more precisely what it means to measure noise in the case of spin ice. The magnetometer will detect a magnetic field which changes with time, as the monopoles hop around. If you look at that noisy fluctuating signal you will find that it will have contributions from all different frequencies of the time-varying magnetic field. If you look at the contribution from each frequency of the magnetic field, you will identify the colour of noise of the crystal's magnetic field. If there were an equal contribution to the signal from all frequencies, the magnetic field would take the form of white noise.

By extending Einstein's analysis it is possible to prove that any paramagnet is expected to show perfect red noise at high frequencies. In an instance of true universality, perfect red noise is also generated by semiconductors as electron–hole pairs pop into and out of existence. So what about spin ice? Here's where Fran's computer simulations came in. She found that in spin ice, at high frequencies, the noise is not red. While it depends on temperature, it is always pinky-red, becoming pinker as temperature increases. The long-range correlations cause the spin ice to behave in a measurably different way from a paramagnet. This implies the spin ice also behaves differently from a semiconductor. In fact that's correct and in accordance with the theory: since elementary magnetic monopoles do not exist, a line of magnetic flux must connect a monopole to an antimonopole, which is not true of an electron and hole. This observation was first made by Paul Dirac in 1931, in the original paper theorising the existence of magnetic monopoles.[13]

A few months after our numerical predictions, Fran gave a talk on the work at a conference. In the audience was Professor J. C.

Séamus Davis. While we had envisaged using a tiny, nanoscale magnetometer, Séamus realised the noise prediction should survive on
everyday scales, into our middle realm. This opened the possibility of
easier experiments. Now, 'easy' in experimental physics might usually mean that something might take a few years with existing technology. But this case proved an exception owing to some incredible
experimental skills by Séamus's then PhD student Dr Ritika Dusad.
A few days after Fran's talk we received an email saying Ritika had
built the experiment, carried out the measurements and confirmed
Fran's predictions: the noise in the magnetic field was not red, as it
would be for any paramagnet, but pinky-red. The conclusion was
that the spin ice was behaving not as if it was made up of magnetic
dipoles but magnetic monopoles. While the monopoles emerge from
the behaviour of many dipoles, the experiment sees the monopoles,
just as, when an owl emerges from the collective behaviour of many
atoms, you see the owl not the atoms.[14]

Ritika later explained to me that she was surprised at the skills
she called upon as an experimental physicist. Principal among these
was the sewing taught to her by her grandmother: to carry out the
measurement she had carefully threaded a coil of wire around the
rather tiny spin ice crystal six times, then connected this coil to a
very sensitive detector of magnetic flux called a 'superconducting
quantum interference device' (SQUID for short). The SQUID has
a range of frequencies of magnetic field to which it is sensitive,
just as your ears are sensitive to a range of frequencies of sound.
Remarkably, these two overlap well: you can hear sounds from about
20 Hz to 20 kHz, while the SQUID detects magnetic fields from a
few hertz to about 2.5 kHz. This means you can convert the magnetic field SQUID signal into a literal noise by creating a sound with
the same amount of each frequency, allowing you to listen to the
sound of the magnetic monopoles. Ritika took these measurements,
and you really can hear the difference between the SQUID hearing a

paramagnet of magnetic dipoles, and a spin ice of magnetic mono-poles. The trick to finding magnetic monopoles was to listen.

Magnetic monopoles in spin ice are one way that a successor to electronics might be created without using electric charge. They are but one aspect of a larger developing field within condensed matter physics: the casting of spells of division, the technical name for which is fractionalisation.

Fractionalisation

Emergent magnetic monopoles are one example of fractionalisation: emergence has divided the magnetic dipoles into two halves – a frac-tion of what they were. To my mind this is one of the most profound examples of emergence: I mean, surely the one thing which *can't* emerge is something *smaller* than the things from which it emerges! But it turns out it can. Another example that is receiving both inter-est from physicists and funding from the computer industry is 'spin–charge separation'. This occurs when many particles, each with a spin and electric charge, combine to create a mix of two types of emergent quasiparticle – one with just a spin, and the other with just a charge. Called 'spinons' and 'holons', they move independently of one another, and even have different masses. They are sometimes compared to the Cheshire cat in *Alice in Wonderland*, which can separate itself from its grin:

> 'Well! I've often seen a cat without a grin,' thought Alice, 'but a grin without a cat! It's the most curious thing I ever saw in my life!'

The reason spin–charge separation is exciting is that moving spins requires less work and generates less heat than moving charges. Rather than electronics, which move charge, these 'spintronic' devices promise huge increases in efficiency. Using them, spin could be used in essentially all the ways that charge is used now in electronic devices, but with less energy expended and at a smaller scale. This technology is not something that is ten years away – spintronic devices already exist. The International Roadmap for Devices and Systems, developed by semiconductor industry leaders, identifies spintronics as a currently viable technology; spintronic computer memory has already been implemented commercially, albeit in a niche position while the technology develops further.

Spin–charge separation occurs in materials which can be thought of as approximately one-dimensional. That might sound strange, but it's not so weird: while real materials are three-dimensional, some have strong bonds between atoms along chains but weak bonding between the chains. An electron trying to move around within such materials then finds it easy to move back and forth along a one-dimensional line, but prohibitively difficult to move off the line. Strontium cuprate $(SrCuO_2)$ is an example of a material where this phenomenon has been reported.[15] Perhaps most remarkable of all is that spin–charge separation is believed to occur *generically* for one-dimensional metals. The basic picture of how it occurs is not so tricky to understand. It was explained to me by Professor Fabian Essler in Oxford, an expert on one-dimensional matter. I will paraphrase his explanation of the phenomenon here.

Empty the pockets of your cloak onto the table of your local tavern; push aside the assorted trinkets, gems and curses scrawled on ancient parchment, and locate all the coins you can find. They will, of course, be rough-hewn groats. Now space these coins evenly along a line, alternating heads and tails. Each coin represents a negatively charged electron; heads will correspond to the electron having spin-up (north pole pointing out of the table), while tails is spin-down (north pole into the table). This is shown in the top line of Figure 29a on the following page.

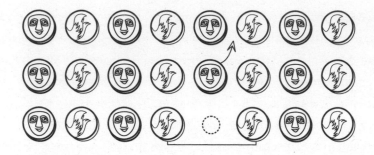

29a: Coins represent electrons, where heads and tails represent spins pointing into and out of the page. Removing a coin (electron) creates a hole, but also a double-tails either side of the hole. This acts as a concentrated region of into-the-page spin.

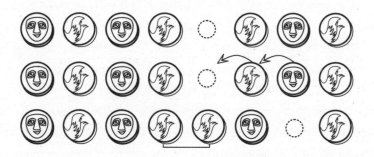

29b: Moving coins into the hole from the right moves the hole right, separating the hole from the double-tails.

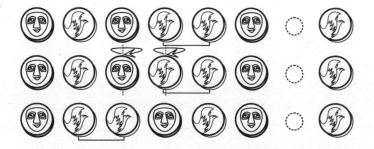

29c: The double-tails can hop by flipping two neighbouring coins.

Now remove a coin from somewhere in the middle (shown in the lower two lines of Figure 29b). This happens, for example, if an energetic photon arrives in the material, kicking out an electron. Say the coin was heads: you now have a hole where the coin used to be, which represents a positively charged hole in the material. But you also have a concentrated region of 'tails', or spin. Here's the cool bit: you can move the charge and the spin separately. To move the charge, perform the trick of the wizards' hats from Chapter V: move the coin immediately to the right into the hole, with the effect of moving the hole to the right. Do that move a couple more times to get the positive charge – the holon – out of the way (Figure 29b). Notice that this leaves the double-tails together. Now, to move the double-tails – the spinon – imagine flipping the left coin of the double-tails to heads. The double-tails has been removed at the expense of creating a double-heads to the left of it. While this gives the right idea, what happens in the real material is that both the left-tails *and* the heads to the left of it will flip: in this way the double-tails becomes another double-tails, moving two places to the left (shown in Figure 29c). This is a less dramatic change to the magnetic field than flipping just one spin, and requires less energy.

By separating spin from charge it is possible to have a flow of spin by itself. To be clear, this fractionalisation takes a different form from that of creating magnetic monopoles. In monopoles the spell of division is to separate the north pole from the south. Here the spell separates the spin from the charge, but the spin still has both of its poles.

While there is much still to understand, fractionalisation has already provided a wide variety of practical uses.

Practical magic

One benefit we can always expect from fundamental science is that the experimental advances needed to study the theories lead to technological developments that benefit society as a whole. It is in the nature of fundamental scientific enquiry that it drives forwards the limits of human understanding.

One possible application of the ideas presented in this chapter is the development of magnetic resonance imaging (MRI).* MRI scanners in hospitals are huge, very expensive devices. Receiving an MRI scan is not pleasant: you have to lie in a big tube while a loud clunking noise occurs around you. The scanner measures the location of water and fat in the body, inferred from the locations of the spins in the atomic nuclei. To detect the spins a magnetic field must be applied. This field has to be huge, because it only couples very weakly to its targets, which is the reason for the expense, intrusiveness and power consumption. The size of the applied field is dictated by the sensitivity of the measurement device. Now, the measurement device Ritika Dusad built to measure magnetic monopoles constituted the most sensitive detection of magnetic flux ever performed. If the detector is more sensitive, the magnetic field employed need not be so powerful. Some of that huge expense and intrusiveness of MRI could disappear, saving hospitals time, energy and cost, and saving patients an often upsetting experience. A paper in the *Proceedings of the National Academy of Sciences* already demonstrated the practical possibility of small-field MRI experimentally.[16]

Electricity and magnetism are ubiquitous in modern technology, but it is fair to say that electricity is the more ubiquitous of the two: we have electricity on tap in our homes but the same is not true of magnetism; magnetic currents don't exist because magnetic monopoles don't exist. If you tried to pass magnetism down a wire,

---

* Scientists, by the way, call this nuclear magnetic resonance (NMR), because it involves the measurement of the magnetic spins of the nuclei within atoms. But it was decided that having the word 'nuclear' in the name of a device you're asked to stick your head into might put people off (plus 'give this person an NMR', when said aloud, might be misunderstood in a hospital).

wherever a north pole goes a south pole follows so as to cancel it out. Harnessing monopoles in spin ices could redress the balance. Professor Steven Bramwell, one of the original discoverers of spin ices, proposed the name 'magnetricity' for this. He and his collaborators subsequently found a good raft of evidence for monopoles by thinking of them flowing as magnetricity through spin ice.

Now, these emergent magnetic monopoles can only exist within spin ice crystals, and we're probably not going to start building power lines out of spin ices and cooling them down to below 2 K, colder than the universe itself. But we might potentially start incorporating the technology into small spintronic devices which are already being developed. Magnetricity promises the opportunity to create magnetic versions of any electronic component that can operate with alternating current. This could be a huge benefit to the widespread adoption of spintronics.

Splitting the difference

Returning to the motivation at the start of this chapter: might fractionalisation provide a fundamentally new approach to computing as we head into a post-Moore's law world? I think there is a good chance that it will. One application of emergent magnetic monopoles that has already been realised is *artificial* spin ices, arrays of magnets each around a thousandth of a millimetre long. As in spin ice, the lowest-energy arrangement has two magnets pointing in, and two pointing out, wherever they meet. Logic-enacting components have already been manufactured in artificial spin ices. As with spintronics, computation with artificial spin ices can be much more efficient than by transferring charge. In fact, artificial spin ices have been shown to be able to operate at the Landauer limit, that fundamental maximum efficiency set by the second law of thermodynamics.[17] Spin ices also forced a more nuanced understanding of the third law of thermodynamics. Recall this states that a perfect crystal must be perfectly ordered at absolute zero, suggesting a single microstate compatible with the lowest-energy macrostate. But spin ice, and ice before it, have about as many lowest-energy arrangements as there are atoms in the crystal. Having many possible arrangements with the same energy is the very definition of disorder: many microstates compatible with the same macrostate, meaning lots of entropy. This realisation required the third law to be polished a little. The modern understanding is this: as the temperature approaches absolute zero, matter need only tend to a state of constant disorder, meaning that the entropy tends to a constant value rather than zero; it need not lose its disorder altogether. Spin ices, by pushing scientists to understand how they exist within the laws of thermodynamics, gave a clearer understanding of those laws themselves.

Emergent magnetic monopoles in spin ices, and fractionalisation more widely, are one possible spell of division we might cast to separate the idea of computers from their implementation in semiconductors. The electronics industry began with quantum mechanics, when condensed matter physicists first understood the semiconductor, and it will end with quantum mechanics: the fundamental limits being reached by semiconductor technology are now on such a small scale that they are quantum in nature. With small enough electronic components, electrons become unruly, tunnelling out of where we'd like them to be.

But rather than try to swim against the current of quantum mechanics, we should try to go with it. Embracing quantum effects promises computers powerful beyond current imagination. This chapter considered the development of condensed matter physics in the twentieth century right up to the cutting edge and its immediate future. Now it is time to look further ahead: to the biggest industry of 2035, which does not yet exist.

VII

Spells of Protection

For want of a better phrase, the birth cycle of the knot masters resembles reincarnation; a true understanding of the process requires some comprehension of the world-view of the knot-making clan, which is influenced by the most important, most mystical and least-well understood of their abilities: their intimate knowledge of both the past *and of the future*. This understanding is tied to the operation of the nexus. While the output of a calculation may be fixed, the intermediate process holds open an infinitude of possibilities.

From these islanders' perspectives, time is not a linear string connecting past to future, but an interwoven web. History is not statically recorded, but is created interactively in the mind of the knot maker as the nexus is dynamically read out. Whereas the smudge of a single written word might alter the meaning of an entire text, or a slip of the tongue might similarly alter a verbal history, the information encoded in the nexus is only able to change if a knot is tied or untied ...

Welcome to the world of tomorrow!

Can you imagine a new technology that the people of tomorrow cannot imagine living without? Here's my answer: quantum computers. In 1985 physicist Richard Feynman made the following observation: there are physical processes known to be impossible to simulate on a computer in any reasonable time; yet reality constantly simulates these processes – by doing them. A horse is a perfect quantum simulation of a horse, accurate from the microscale to the macroscale and encompassing all its emergent properties along the way. Therefore, Feynman reasoned, if computers employed quantum mechanics they would be able to do certain useful calculations much faster than is presently possible.

These quantum computers would have profound applications. They could accurately deduce the behaviours of elementary particles, a task presently reserved for huge particle accelerators such as the Large Hadron Collider. They might have applications in biology and medicine: for example, they could trivialise genome sequencing, helping to fight emerging viruses. They could be used for drug prediction, discovery and even design at the molecular level. They might have applications in chemistry, such as designing better batteries, vital to reducing our reliance on fossil fuels. They could simulate molecules, reaction rates and predict new methods of synthesis: the production of ammonia, employed as a fertiliser for crops the world over, uses around 2% of the world's energy; yet bacteria can produce ammonia far more efficiently, and molecular simulation on a quantum computer could reveal how.

Quantum computers are no pipe dream. In fact, they already exist: in October 2019 researchers at Google published evidence that their quantum computer had carried out a calculation more than three million times faster than would be possible on the world's fastest supercomputer. In December 2020, a group in Hefei in China used a quantum computer to solve a problem in 20 seconds that would take 600 million years classically.

But there is a problem: the seeming impossibility of achieving *scalable* quantum computing. As if stuck in a spider's web, for each inch of progress we wriggle we simply find ourselves more stuck. This is because the source of quantum computing's power is exactly why it is hard to scale it up.

Scaling is key to practical application. For example, it is often said that spider silk is stronger than steel. If so, why do we continue to make things out of steel? The answer is that the strength of spider silk does not scale. It derives from the bonding between water molecules on the microscopic scale, and it is only strong when microscopically thin. Thicken the silk and the water molecules stay the same size, so thick spiderwebs would be hopelessly weak. Through incredible feats of engineering we now have a quantum computer which is to classical computers as spider silk is to steel: far superior, but only on tiny scales. As it stands, scaling quantum computing for use on practical problems of everyday scales appears more unachievable than scaling spider silk.

My friend and former colleague Steven Simon, Professor of Theoretical Condensed Matter Physics in Oxford, is one of the world's leading experts on quantum computing. As he puts it, quantum processes must be free from noise, meaning their environments must be very cold, and very clean. Minimising noise is an engineering challenge and there have been impressive advances, but progress becomes exponentially harder at each step. However, Steve notes that there may be another possibility: to learn to become deaf to the noise. This is the path of theoretical physics.

What we need is some method of protecting the quantum information from the destructive effects of the outside world. The spell of protection we will weave is called *topology*: the study of shapes, in a general sense, such as the tying of knots and the perforation of holes. It is an art more ancient than writing; to learn this spell will require us to question some notions of reality we hold sacrosanct. Let us first take stock, and assess the difficulty of practical quantum computing.

The garden of forking paths

A classical computer, such as you have in your mobile phone, stores information as bits, 0 or 1. The power of a classical computer is proportional to the number of bits it can hold in its memory: to double the power you double the number of bits. I distinctly recall my surprise at learning that supercomputers are simply huge racks of ordinary computers wired together: I learnt this from an experimentalist who found that the most efficient way to buy computing power was to buy a vast quantity of second-hand Playstation 2 consoles.

A quantum computer, on the other hand, stores information as quantum bits, qubits. These are quantum superpositions of 0 and 1. It's the ultimate form of parallel processing: the universe itself holding open the possibilities of an unmeasured quantum system. Contrary to the classical computer, the power of a quantum computer increases *exponentially* with the number of qubits: to double the power, you simply add a *single* qubit. But there's a catch: adding each additional qubit is exponentially more difficult, for each new qubit must combine with all others.

This is a shame, as quantum computers could carry out certain calculations massively faster than their classical counterparts. A calculation is encoded in a computer as an *algorithm*. These are often compared to recipes: a sequence of instructions an alchemist, say, must follow to produce a desired outcome.

The first quantum algorithm was devised by Professor David Deutsch in 1985, and it was developed in subsequent years into what is now called the Deutsch–Jozsa algorithm. The calculation it performs was chosen to be easy for a quantum computer and prohibitively hard for a classical computer. In his book *The Fabric of Reality* Deutsch instead highlights the example of Shor's algorithm, the first quantum algorithm designed with a practical application in mind. Shor's algorithm is a procedure for using a quantum computer to find the set of prime numbers which multiply to give a chosen number.* If Shor's algorithm could be implemented it would have major implications for internet security. The RSA cryptographic standard used to secure most internet communication, from emails to bank transfers, relies on the practical impossibility of factoring large numbers into primes. Every method we know takes a prohibitively long time, and the numbers used for internet security are hundreds of digits in length. It is impossible to prove that no fast method exists; but it has been proven that if such a method were found, many other very hard problems would become easy. Since these problems have collectively remained unsolved for a very long time (in many cases hundreds of years), cryptographers have judged it a safe bet that they will remain unsolved forever. Yet quantum computers could solve them in the blink of an eye.

Deutsch raises an interesting question: where does Shor's algorithm derive the power to operate so much faster than any possible classical method? His answer is that the power comes from parallel universes. Deutsch is a leading advocate of the *many worlds* interpretation of quantum mechanics, according to which, whenever a quantum superposition is measured, the universe forks into multiple parallel universes, and each outcome of the measurement is obtained in one universe. These universes then live inside a larger multiverse; this idea has inspired many great works of fiction, from *Back to the Future*, to the Marvel universe, to the mid-1990s television series *Sliders*. The idea was even prefigured in fiction: Jorge Luis Borges's 1941 short

---

* I recall being surprised when I first learnt that any integer number can be written as a product of primes – but this has to be the case, because if a number can't be written as a product of other numbers, that means it is prime itself.

story 'The Garden of Forking Paths' concerns a book of that name, written by one of the characters, in which whenever a character faces a decision the story follows all the possible choices. Borges's story is quoted by Bryce DeWitt, who gave the interpretation its name, in his 1973 book on the subject. However, for now at least, the many-worlds interpretation remains a question of personal belief rather than physics. This is true of all interpretations of quantum mechanics: since they all agree with the mathematical predictions of the theory, they agree with one another about the outcomes of any conceivable experiment.

So then where *does* the power come from? An uncontroversial answer is that it comes from whatever makes the quantum world quantum. What is it that distinguishes the quantum world from the classical middle realm? Essentially, it is two things. The first was the subject of Chapter V: the ability of quantum particles to exist in superpositions, combined with the fact that when measured they are always found in one outcome. The second is a property entirely without everyday precedent. It is called quantum entanglement, and it is the most magical property of the universe that I know of.

Before going into what entanglement means, allow me to motivate it in terms of its practical application to quantum computers. The opening fictional passage of this chapter can be taken as a reimagining of some of the key methods of a quantum computer. The easiest way to picture what a quantum computer does is to think in terms of 'quantum circuits', which are used to visualise and design quantum algorithms. A quantum circuit is a set of parallel horizontal lines, one for each qubit, like a musical stave or unwoven strands of a string combed out straight from left to right. An example is shown in Figure 30.* Like sheet music, strands are

---

* This quantum circuit enacts a Fourier transform for the right set of operations, although the details and meaning of the symbols are not important here.

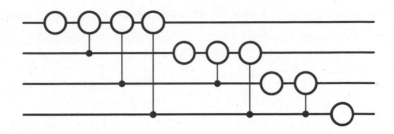

30: A quantum circuit. Each strand represents a qubit. The first operation, read from the left, acts on a single strand, while the second acts on two strands. In general an operation on multiple strands can create entanglement between qubits.

read left to right, and all strands are read simultaneously (just as you might play multiple notes simultaneously on some instruments). Each qubit – each strand – starts out in a known state, o or 1. Reading along the strings, there are two types of operation which can be performed. The first acts on individual strands, in much the same way that a single note can be played in music. This might create a quantum superposition of o and 1 along that strand, for example. Superposition is one element of quantum mechanics we have already seen, and it is a key process in quantum computation. However, there is a second type of operation which combines multiple strands. This is indicated by some symbol linking multiple strands, and you might think of it as like playing multiple notes simultaneously (a chord). The fates of these strands are subsequently connected. Intuitively, as the name suggests, these qubits can be 'entangled' by such operations. Finally, the right-hand ends of all the strands represent the end of the calculation. At this point the qubits are measured; they will again take definite values, o or 1. The fact that they give definite outcomes when measured, despite necessarily being indeterminate in the middle of the calculation, is part of the magic of quantum mechanics.

Our focus in this chapter is on those operations which entangle multiple strings.*

Is the moon there when nobody looks?

The title of this section is ~~shamelessly stolen from~~ *a homage to* a popular article on quantum entanglement by David Mermin.[18] The opening line of that article is characteristically to the point:

Quantum Mechanics is Magic.

I previously introduced Mermin as a master of quantum mechanics. In fact he is a renowned condensed matter physicist who is equally well known for his clear explanations of the most magical aspects of the quantum world.† His textbook on condensed matter physics has been the go-to reference for every undergraduate student since the mid-1970s. I think it is significant that he leads his article with the quote above. When you work as a professional quantum mechanic it's tempting to deny the magic, as if admitting that there are things we don't understand is to confess some personal failure. You were so enthralled by quantum mechanics when you were young that you became a physicist to spend more time with it, but at some

---

* By the way, the fact that there are the same numbers of string ends on the left and right might be intuitive, but it's actually quite subtle. Recall that a traditional piece of logic might take in two statements and give one output: IF (A AND B) THEN C would have two lines in, A and B, and one line out, C. A transistor would enact such logic. But here the number of lines is always fixed, meaning any process can be reversed. All quantum computers are reversible, but classical computers can also be made reversible, and this can circumvent the Landauer limit on maximum efficiency imposed by the laws of thermodynamics.

† That Mermin's name is one letter away from Merlin is purely coincidental.

point you gave in to the temptation to say the magic's gone for you. But then you're stuck in stage two: you've worked out how some of the magician's tricks work, and have declared them boring. But as the magician taught me in the desert, if real magicians are excited by magic, so should you be. Mermin's quote puts this succinctly, and his authority enables other physicists to admit they see the magic too.

Mermin was born in 1935, the year Einstein wrote the first paper on quantum entanglement. In an interview he stated:

What first drew me to physics was magic. It came in two varieties: relativity and quantum mechanics.

That is, Einstein's work of 1905. I once met Mermin, but to my shame I did so without knowing who he was. I was a first-year undergraduate, and I'd signed up to a quantum mechanics conference in Germany organised by Professor Dr Anton Zeilinger, who led the world's first 'quantum teleportation' experiment. I didn't know anyone who'd been to a scientific conference before, and I thought I might be found out as too junior and booted out. Adding to this, my arrival had been slightly unconventional. I turned up a day early, imagining I'd use my broken German to find a guest house – but nowhere was open. The conference was being held at the headquarters of the German Physical Society, an eighteenth-century palace. I dared not turn up early in case they discovered my imagined fraud, so, instead, I curled up to sleep in a nearby building site, wrapping myself in my robe-like coat. At some point a rat ran past me, prompting some re-analysis: I decided it was a better plan to sneak into the palace and find somewhere to hide. Getting in, I found my way to a library and went to sleep in a wing chair by a warm fireplace; I put a book on my lap, so that, if questioned by a footman, I could pretend to have simply fallen asleep reading. At 7 a.m. I decided it was late enough to check in: 7 a.m. was just supernaturally business-like, not dangerously eccentric. Heading to the front desk I found nobody there, so I located the key to my room on a peg. Letting myself in, I found the room contained two beds – and that I had a roommate! When he awoke he explained that he'd been there the whole previous day: we were supposed to let ourselves in. Didn't I get the memo?

The rest of the conference went much more smoothly. At some point I found myself talking to Mermin. Aside from becoming thoroughly acquainted with his textbook the following year (to such a degree that I could consistently open it on precisely the page I needed for a variety of topics) I subsequently found myself consulting him on research problems as my career progressed. He has always proven as helpful and encouraging as he was during our first meeting in the palace.

It was from an article of Mermin's that I first felt I'd really started to understand quantum entanglement. I will reproduce his example here, rephrased only in context.

Y golchwyr nos

In 1935 Einstein published a paper with Boris Podolski and Nathan Rosen intended to show that quantum mechanics could not be a full account of reality.[19] It outlined what is now called the EPR paradox, after its authors' initials. As Mermin put it: 'The EPR experiment is as close to magic as any physical phenomenon I know of, and magic should be enjoyed.' The EPR paper took aim at the idea that when you measure certain quantum systems they have a 50–50 chance of giving each of two outcomes, like tossing a coin and slapping it down on the table before looking. As we saw in Chapter V, when tossing a quantum coin the system apparently only chooses heads or tails at the moment you look. There is an element of interpretation in what I just said – the many worlds interpretation would suggest the heads and tails branches of the universe stop talking to one another when the measurement is made, but that the multiverse otherwise continues unabated. What's agreed on is that there is a measurable difference between the quantum coin and a classical one. Remember the electron in the atom: if it had a well-defined position before you measured, it would fall into the nucleus. So something strange must happen in the process of measurement, however you interpret the mathematics. Einstein quite reasonably took objection to this. As a friend of his recounted:

> I recall that during one walk Einstein suddenly stopped, turned to me and asked whether I really believed that the moon exists only when I look at it.
>
> A. Pais, *Einstein and the Quantum Theory*

The remarkable thing about the EPR paper is that it led to an experimental test which could actually distinguish between the two cases – whether the quantum coin has an outcome before being measured, or not.

Einstein's argument, that it has an outcome, uses the property of quantum entanglement, which is when the outcome of a measurement of one quantum particle implies the outcomes of measurements of other particles. Before I give Mermin's argument, allow me to summarise the essence of the magic of the EPR paradox. The entangled particles could be very far apart when one is measured; measuring one particle would seem to dictate the outcome of a distant measurement instantaneously, but that would go against one of the central ideas of special relativity – that nothing can travel faster than light. To understand how to resolve this apparent contradiction, let us relocate to a more fantastical setting.

Celtic folklore warns that lonely travellers on weary nights may chance upon three night-washers – *y golchwyr nos*, in the ancient tongue – cleaning the shrouds of the dead. If the washers see you they will force you to help them wash clothes. But be careful: for if you wring the same way as them you will be pulled into the clothes and killed. If you wring oppositely, the night-washers will instead grant you three wishes.

Walking this way and that along moonlit branching woodland paths, you and two friends find the night-washers by a dark pool. Being well versed in the ancient ways you wring oppositely; and being well versed in logic you make the following wishes. First, you wish them to speak only the truth. Second, you wish them to tell you the best wish. And third, on the advice of the second, you wish to learn the world's most powerful magic.

They agree. The first washer plucks the moon from the sky as a cold hard marble and passes the marble-moon to the second washer, who tosses it from their left hand to their right. They somehow now have marbles in both hands. They toss each to their fellow washers; before you can wonder how the washers now have one marble each or who put the moon back in the sky, each washer has taken one of your trio by the hand. They stand back-to-back, you three in a ring around them. Each of you now has a different washer before you. Each washer holds their closed hands in front of them, each with a marble in one hand, and asks the person in front of them to guess which hand the marble is in, on a count of three. They declare that their magic will be of the following form.

   i. Whenever one of you picks the left hand, an odd number
       of you will be correct.
   ii. Whenever three of you pick the left hand, an even number
       of you will be correct.

You play many times, and the night-washers are always correct: whenever one or three of you pick the left hand, i and ii are obeyed.

Your appreciation passes through two stages. First, that's magic. How did they know? Second, you begin to rationalise. At least one of them must be changing which hand their marble is in through sleight of hand. Otherwise there would always be a 50–50 chance of an odd or even number of you being correct. Furthermore, you deduce, at least one of them must be listening to what all three of you choose.

The washers see that you're stuck in stage two, but they are contractually obliged to take you to stage three; so each washer takes their human partner to a separate tower, each of which is a day's raven flight from both others. At midnight every night a raven arrives at the window of your tower with a marble. Each night your washer takes the marble and asks you to guess which hand it is in. Fashioning a quill from a raven feather, each day you note down which hand you chose and whether you were correct. After many months of this daily routine you travel back to that fated pool to meet your friends. They too have raven-quill lists. Comparing lists, you can't believe it: every evening, conditions i and ii were always obeyed.

And now you reach stage three, because your washer can't possibly have known what your two friends chose. Even if they were using sleight of hand to decide whether or not you were correct each time, they cannot have known what outcome to choose to make i and ii be true. The washers can't have communicated, because nothing travels faster than a raven. OK, you think, but maybe there was a secret message carved in the marble. Or maybe the washers use a preset pattern of whether they reveal a marble or not. Or maybe they coordinate their responses using the colour of the sky or the temperament of the moon. But none of these will help them win, for the following reason.

If there's a secret rule it must specify the outcome for both hand choices for each person, because you get to choose your hand freely. For example, a secret rule compatible with i has each washer reveal a marble regardless of which hand each person chooses. Whenever one person chooses left, three people are correct, because three people are always correct in this case: three is odd so this matches i. But that secret rule is not compatible with ii, because if three people choose left, then three people are still always correct. Three is not even, so this doesn't match ii.

With a bit of thought you can work out every possible secret rule compatible with i: there are eight, and I explain how to get them in the Appendix (see page 295). However, ii is violated for all eight! No matter what plan the washers might have agreed upon beforehand, and no matter what information is hidden within the marble, the sky or the temperament of the moon, there is no possible way for i and ii to be true unless the answer of one washer (yours, say) changes based on what your friends chose. But that would need them to know the other choices faster than a raven.

I recommend you try to think up any possible way i and ii can always be true, while still permitting you and your friends free will. You can think up any mechanism you like; the only restriction is that nothing communicates faster than a raven. The more you try to beat the game the more convinced you will be that doing so is impossible.

Have you tried? And you agree i and ii can never be obeyed consistently? You're convinced: if that happened, it would be magic. Well here's the thing: when physicists did the experiment with entangled particles, conditions i and ii were always obeyed.

Stark raven lunacy

Rather than three ravens delivering marbles to three distant towers, the real experiment has three entangled photons arriving at distant detectors.[20] Actually the detectors are only a metre or so apart, but the measurement at each detector can be recorded fast enough that light would not have time to travel from one to another. Since nothing can travel faster than light, there is no way for the result at one detector to be communicated to the others before they detect. Rather than choosing the hand of a night-washer, each detector chooses to measure the polarisation of light in either a horizontal or vertical direction. The detector is programmed to make this choice randomly just before the photon arrives, so the other detectors can't know which it chose. The two outcomes are whether the photon has the chosen polarisation. Conditions i and ii are measured to always be true when the photons are described by a certain entangled wavefunction, called the GHZ state (Z for Zeilinger, who invited me to the palace on that fateful night). The experiment implies the result that Einstein found so problematic. The outcome of the measurement cannot be predetermined: the value of the quantum coin is neither heads nor tails before it is measured. If it were, you'd be back to the eight secret rules above (which you have already agreed cannot work). These secret rules are called 'local hidden variables' in quantum mechanics; this experiment shows that they are incompatible with our universe. Probabilities in quantum mechanics do not just quantify our personal lack of knowledge: they must be something more profound.

It is tempting to think that this experiment could be used to communicate faster than ravens, but that is not the case: all you experience is the washer in front of you revealing whether or not the marble is in the hand you chose. In the real experiment it turns out that you're correct exactly half the time, regardless of your choice. It is only when you travel back to meet your friends that you see your results were magically correlated.

No accurate classical analogy for quantum entanglement is possible. But if we look instead to the *impossible* there is a sense in which something similar is intuitive, albeit totally wrong: jinxing. Say you're waiting for the outcome of a job interview, and someone asks you how it went. You feel it went very well; but don't you get a strong urge not to say it went too well, in case you jinx it and don't get the job? There's no causal mechanism by which your positive response could actually make that happen, but it's a bit of magical thinking we all engage in from time to time. If I try to pin down what form my fear of jinxing takes, I suppose it's that until the outcome is told to me, I feel that I live simultaneously in two possible worlds, and if I say out loud that I think I got the job, then I seal my fate as living in the world in which I didn't get the job, even though the decision is made far away from me, in my past or future. Measuring a particle can similarly be thought of as sealing the fate of its entangled partner, even if that partner is far away and measurements on it were made in the past; the two must give consistent results even though their fates were not determined before measurement.

While you can't use magic to communicate faster than ravens, it is far from useless. If the world were classical, the night-washers could only be correct half the time. But with access to entangled quantum particles they can be correct every time. This has already been put to practical use in 'quantum cryptography'. The basic idea is that if the raven is intercepted and the marble inspected along the way, the outcome is measured before the marble arrives. The entanglement with the other marbles is lost, and conditions i and ii can no longer be obeyed. By detecting the lack of entanglement in your arriving marble you know you have an eavesdropper, and stop communicating. The first bank transfer to employ entangled photons for quantum cryptography occurred in 2004; in 2017 entangled photons were successfully bounced off satellites for long-distance communication.[21] I gained my first taste of quantum entanglement from Brian Greene's inspirational book *The Fabric of the Cosmos*. His description of entanglement experiments blew my mind: I could smell on them the thick scent of magic. When Greene wrote his book in 2004 the latest experiments involved entangling pairs of particles. Yet my newly repaired mind was blown anew when, a few years ago, it came to be understood that there are states of matter in which *all* the quasiparticles are entangled with all others. They're riddled with entanglement; defined by it, even. Entanglement can survive the thermodynamic limit! And this is fitting, as entanglement is arguably *the* emergent property: you literally can't separate the sum into parts!

If you're keen to hold a lump of entangled stuff in your hand, you can. In fact you regularly do: all matter is governed by quantum mechanics, and all is entangled to some degree. But that's a bit of a cheat: quantum mechanics is magical precisely because we don't see its more fantastical effects in our everyday world. Now, wizards are practical people: sure, everything is quantum – but what is *practically* quantum?

A *coherent philosophy*

Knowledge of the word was not sought, it was stumbled upon.

The moon takes its rests in those instants when no one is watching. Calabash had meant to look down, but momentarily hesitated, and in a glimpse found the word written across the landscape, in the shapes of the shadows and leaves and clouds on the horizon. No one had seen the word before, and no one would see it again, but there, in that moment, Calabash saw it written clear as markings on a page. Not caring for knowledge, he shouted his word into the forest to be rid of it. But the word was not gone, it was merely shared out amongst the plants and trees and creatures and water and birds in the sky. And Veryan talked to the trees, and to the butterflies, and the snakes on the ground, and the bubbles in the stream, and slowly, methodically, she pieced the word together. And when she was done, she did not know the answers, but she knew where she must look to find them.

Thermodynamics introduced the idea of separating the *system* (the thing being studied) from the *environment* (everything else). Nowhere is this as important as in quantum mechanics. Here, the system, say a lump of matter, behaves according to the Schrödinger equation, which says that its particles become increasingly entangled with time. The environment can be thought of as 'measuring' the system, in the sense that it interacts with it and becomes entangled with it. That was the original idea of entanglement, when it was put forth by Schrödinger in a letter to Einstein: he envisaged an experimental detector becoming entangled with the quantum system being studied, their fates entwined. Measuring which hand the marble is in gives a definite outcome, despite the quantum weirdness going on beforehand; and when the environment measures the system it seemingly leads to a similar state of certainty. The resulting loss of apparent quantum weirdness is called decoherence.

If someone is speaking coherently they are making sense. A wizard, of course, will often do the opposite, mumbling incoherently to themselves as they focus their minds on lofty matters. Coherence gets a technical meaning in quantum mechanics, but one that maintains this intuition. In essence, when a quantum system is coherent it is able to work its magic. On the other hand, the constant interaction between a system and its environment leads to decoherence. The standard story is that this is how the middle realm emerges from the quantum, and how matter exists: it is why our daily lives aren't plagued with quantum weirdness, and why, when you return home from a long day of spellcasting and rest your staff in the umbrella stand by the front door, you can rely on it not tunnelling through the wall into your neighbour's tower. The moon *is* there when nobody looks, the story goes, but not for the obvious reason.

Except there's a problem with this. Decoherence doesn't actually eliminate quantum effects, it just spreads them out. The environment can't literally measure the system, because the environment is itself a quantum system. Whatever you pick to be your system will decohere, with time, with its environment. Its quantum information slips into the surroundings as the vibrations of a trapped fly pass into a spider's web. But like that word shouted into the forest by Mr Calabash, it is never really gone: it is just hard to piece together again. The situation has a close analogy in thermodynamics: while energy is conserved overall, a particular system may not conserve energy. The swing of the hypnotist's watch lessens with time: the energy of the swing passes to the environment as vibrations and heat. It may be useless to us, but it's there. Like energy, quantum information is conserved, and never truly lost.

So the magic persists. Until we understand quantum measurement we cannot say for certain whether the moon is there when nobody looks. This 'measurement paradox', as it is known, is one of the biggest open questions in physics and philosophy. But decoherence explains the *apparent* loss of quantum weirdness when many particles get together. The weirdness spreads out and we lose track of it. It is this which halts the progress of anyone attempting to bring the practical benefits of quantum mechanics to our middle realm: each additional qubit added to a quantum computer increases the computing power exponentially, and it would only take about 270 qubits to simulate a universe containing as many particles as our own, if those particles behaved classically. But adding each qubit also exponentially increases the difficulty of adding the next. Each additional qubit must remain coherent with all previous qubits, and entanglement loves nothing more than to spread. Worse, it is not clear that the maximum number of qubits we will ever achieve through the perfection of this art will be useful. Moore's law may be ending for traditional computers, but they can already do useful things.

We need a spell to protect the entanglement from spreading out, so that we can continue towards those possible futures in which scalable quantum computing is a reality. To reach these futures we can look to the far past: to the ancient art of knot tying.

Much ado about knotting

There are hints from cultures around the world that knots predate writing as the original form of record keeping. The best-known examples are *khipu* (or *quipu*), numerical tallies stored as sequences of knots by the Inca and other cultures in the Andes that inspired aspects of what Veryan is reading in the library. Some academics have argued that certain knots in *khipu* acted as a regional identifier, meaning that knots were used to record language. In the *I Ching*, an ancient Chinese divination text dating to around 1000 BCE, knots are mentioned as recording language; needles made from bone, believed to have been used for knotting, have been dated to 100,000 BCE.

Knot magic falls into three broad categories: wind magic, health magic and love magic. A knot maker might have been visited by sailors embarking upon ocean crossings to create a wind knot, three identical knots tied along a line. The knot maker would enchant each knot with breath, spit and words, to bind in it the power of the wind. Finding their ship adrift, the sailor can untie the first knot to release a calm breeze by which to escape to more suitable conditions. In more severe cases the sailor might untie the second knot, releasing a powerful gale. Only a fool would untie the third knot, as this would release a tempest capable of breaking the mast of the sturdiest of vessels in two. Perhaps the last knot was there as an insurance policy: provided the sailor is too superstitious to use it, unskilled weather-workers could cover up their faults by claiming some miscommunication as to the strengths required of each knot, arguing that their third knot should have been used.

Records of health knots as preventatives and cures date back at least as far as eighth-century BCE Babylonia, surviving as cuneiform incantations pressed into clay. Cyrus Day, in *Quipus and Witches' Knots*, provides examples of the use of health knots from many times and places on Earth, suggesting that 7,000 years would be a conservative estimate of the age of knot magic. For a headache, the knot maker might tie a band around the head, accompanied by the appropriate incantation: as the band is released, so is the headache. Other afflicted parts of the body might be similarly tied and released; alternatively, the ailment might be trapped in the knot, which is taken off intact, tied thoroughly, and discarded far from civilisation (either in a stream close to the sea, or far out in the desert). In preventative spells, the knot maker might advise that all knots in a house be untied when the birth of a child is imminent, in order that the baby not be 'caught': shoelaces should be undone, curtains released and locks unlocked.

Love knots are surprisingly prevalent in modern society. Many marriage ceremonies involve a ritualistic binding of the hands or similar, and wedding rings, being closed loops, similarly symbolise connection and unbreakable bonds. The great multitudes of locks attached to bridges by couples around the world suggest knot magic is still very much believed in.

The history of knots in physics is more recent, but no less eminent. It was inspired by the physical observation that smoke rings, such as a wizard might blow from their pipe, survive all manner of disturbances from the air. They can wobble and stretch, but they never seem to break open, as if protected by some enchantment. The physical mechanism was explained by Lord Kelvin in the nineteenth century. When the wizard blows a smoke ring they are creating an air vortex, like a tornado with its top and bottom connected together to form a ring. The smoke gets caught up in the vortex and lets you see where it goes. Kelvin created a model of this by making some simplifying assumptions about the air, such as that it is entirely lacking in viscosity. Within the model he was able to prove mathematically that if a closed vortex can be set up it must survive forever. In reality air has a little viscosity, which causes smoke rings eventually to dissipate. Nevertheless, they survive long enough that they can be used for some pretty impressive magic: if you cut a ten-centimetre-diameter circular hole in a cardboard box, fill the box with smoke and then whack the sides, you can shoot a smoke ring stable enough to knock over paper cups at a distance of five metres or so.

Mathematical knots are like those you would tie in your shoelace except, like the smoke ring, they have their two ends joined to form a loop. The smoke ring takes the shape of the simplest knot, called the 'unknot' – because it is unknotted. The next simplest knot is the first one you might naturally tie. It is called a trefoil knot, and it appears as a frequent motif in Celtic art: it can be found decorating the eighth-century *Book of Kells*, the eleventh-century Funbo runestones in Sweden and the twentieth-century cover of the album *Led Zeppelin IV*. Inspired by Kelvin's work, his friend the mathematician Peter Guthrie Tait set out to tabulate all possible knots. A sample of his knot table is presented below:

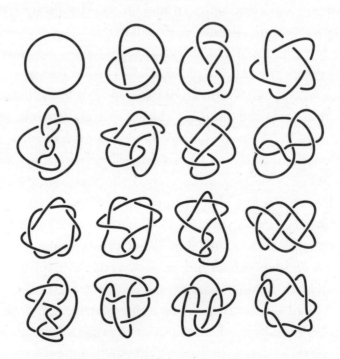

31: The knot table.

The trefoil knot is top row, second from the left. However you draw it, the string of a trefoil knot must cross itself at least three times. This is an example of a 'topological' property. Topology is the mathematical study of shapes; two shapes are considered to have the same topology if they can be deformed into one another without cutting or joining. For example, a wizard's pipe has the same topology as the smoke ring it blows: they both have one hole. Even though they look quite different, the shapes can deform into one other without cutting or joining.

The earliest problem in topology is often said to be the seven bridges of Königsberg (Figure 32). In 1736 the mayor of Gdańsk wrote to the legendary mathematician Leonhard Euler with a puzzle that had been confusing the townsfolk of nearby Königsberg. In that town, seven bridges connected four land masses. The inhabitants walked many paths through their hometown, but one in particular was long-sought and never walked: a path that crossed each bridge exactly once. Finding the path is a question of topology: the bridges could be twisted or stretched, or the islands grown or shrunk, and it wouldn't change the problem. Only if a bridge were broken, or a

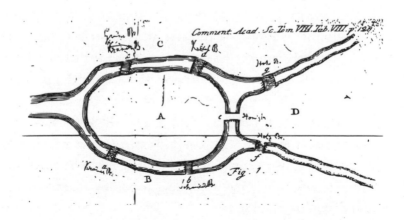

32: Euler's original drawing of the seven bridges of Königsberg.

new bridge added, would the problem change. It is fun to try to solve the problem yourself. However, rather than find a solution, Euler instead proved it is impossible.

An earlier precursor to topology, the problem of the 'knight's tour', dates back at least as far as the ninth century. A knight in chess moves two squares along and one across (or vice versa). The problem is to find a sequence of hops so the knight visits every square on the chessboard exactly once. Euler also provided a number of solutions to this problem. An earlier solution was provided by the Anatolian chess master al-Adli ar-Rumi in 842 CE. Around the same time, a Kashmiri poet called Rudraṭa came up with a truly masterful solution. He wrote a poem in Sanskrit consisting of four rows of eight characters (half a chessboard), each a syllable. The poem can be read in the manner you are reading this, left-to-right and top-to-bottom. But it can also be read by following a knight's tour across the characters. Remarkably, exactly the same poem results.

I once saw a great illustration of topology by Professor Duncan Haldane. Haldane was one of three recipients of the 2016 Nobel Prize in Physics, for 'theoretical discoveries of topological phase transitions and topological phases of matter'. I shared an office with him at a conference in 2015. It struck me that when anyone asked him a question he would look directly at a different person while answering – and the answer he gave always seemed to be to a different question. When illustrating topology he showed a picture of a mug, explaining that it has one hole, through the handle. Then he showed a picture of a two-handled mug: called a 'lovers' mug'. It has a different topology, as it contains two holes, meaning that it can be shared by two people simultaneously. Then he described a *three*-handled mug, which has a different topology again. He called it a 'Californian lovers' mug'.

The essence of topology's spell of protection is that cutting and joining are often much more difficult than bending and twisting. You can twist or jumble or shake a knot, but it will remain the same knot. The unknot will not become a trefoil. Nature has found practical uses for knots: the hagfish, which is shaped like an eel, is known to tie its body in a knot to escape the grasp of predators; the tailorbird uses strands of spiders' silk to sew leaves together to make a nest; DNA chains are sometimes found tied into knots, possibly lending them stability.

In 1997 a Russian physicist, Professor Alexei Kitaev, put forth a remarkable proposal. Imagine a quantum algorithm could be encoded into a knot. Then perhaps it could be protected against decoherence like the smoke ring is protected against buffeting gusts of air. The noise of the world would still be there, but the quantum computer would be deaf to it. This was the first proposal of what would be called a topological quantum computer. Kitaev went on to explain how to tie these quantum knots, using the magic of emergent quasiparticles.

Topological matters

Topology's spell of protection, boiled down to its essence, is this: just like you can't be half pregnant or half in love, you can't have half a hole. While this might seem to only have rather specific applications, topology appears behind a great many practical applications in day-to-day life. One abstract example is provided by that renowned source of ancient wisdom, the hit 1995 film *Die Hard with a Vengeance*. It popularised the following riddle. You have a five-gallon jug, a three-gallon jug and a pond. A megalomaniacal villain with an unconvincing German accent has threatened to detonate a bomb unless you place exactly four gallons of water onto a scale, to within one ounce. You only get one go. How do you do it? After some quick thinking, Bruce Willis and Samuel L. Jackson get to the solution in the nick of time. Fill the five-gallon jug completely. Pour this into the three-gallon jug until it is completely full, leaving two gallons in the five-gallon jug. Empty the three-gallon jug into the pond and pour the remaining two gallons from the five-gallon jug into the now-empty three-gallon jug. Refill the five-gallon jug entirely and use this to fill the three-gallon jug. This takes precisely one gallon of water, leaving precisely four gallons in the five-gallon jug. Yippee-kay-ay, puzzle lovers! They turned a question about guestimating from a continuous range into a sequence of steps in which a jug is either entirely full or entirely empty. The statement 'This is full' is either true or false: it can't be half true, just like you can't have half a hole. They had turned the problem into a question of topology.

A classic example of topology in condensed matter physics is the quantum Hall effect. Recall the Hall effect we met in Chapter V: take a thin sheet of metal longer than it is wide; pass a current along its length and pass a magnetic field through it by bringing a magnet close to it. Connect a voltmeter across the width, and you will detect a voltage caused by the current deflecting sideways; if you increase the magnetic field you will find that the voltage increases proportionately. However, in 1980 Professor Klaus von Klitzing found that if the magnetic field is very large, and the material is very cold and free from defects, the voltage no longer increases continuously. Instead, it jumps by fixed amounts, becoming precisely *quantised*. In fact this is the result of the conductivity of the material itself becoming quantised. Every voltage von Klitzing measured was an exact integer multiple of a smallest amount. You can't have half of one of these jumps, just like you can't have half a hole. This is the 'integer quantum Hall effect'. The jumps can be measured so precisely, in fact, that they came to define the units used to measure them.

The integer quantum Hall effect is topological: von Klitzing's integers are now understood as counting holes in that most mysterious of entities: the quantum wavefunction describing the electrons in the material. The full mathematical explanation is complicated, but it is understood.

What is less well understood is an observation made in 1982 by Dr Daniel Tsui and Dr Horst Störmer: if the material is made even purer, and even colder, the voltages cease to be integer multiples of that smallest amount and instead become *fractional* multiples. One-third, two-fifths, five-halves, that sort of thing. This is the fractional quantum Hall effect; it is a spell we are still learning, but physicists have begun to weave their fables around it.

The appearance of fractions seems to go against the central tenet of topology. What happens when the voltage is one-third of the smallest integer amount, say? Can you now have a third of a hole? You can't. But recall those spells of division, the magic of fractional-isation: when many particles interact, the result can resemble a *fraction* of a single particle. In condensed matter physics you *can* have a third of an *electron*, or at least an emergent quasiparticle resembling it. The fractional quantum Hall effect can again be thought of as counting holes in the wavefunction – but it is no longer the wave-function of the electrons; it is the wavefunction of emergent quasi-particles with one-third their charge.

Many of the most interesting states of matter involve a coor-dination of huge numbers of electrons, moving as if in some care-fully choreographed dance. The fractional quantum Hall effect is just such an example. Magnetic fields cause charged particles such as electrons to move in circles. Dancers move in circles around one another while each dancing around others simultaneously. Different sizes of magnetic field specify the number of steps each dancer takes to complete a circle. This analogy was devised by Professor Xiao-Gang Wen at MIT, who has done seminal work on the theory of the fractional quantum Hall effect. He once explained to me that he takes inspiration from the Daoist idea that the absence of things can be as important as their presence. He gave the example that a room without a door is as useless as a room without walls. It is a perfect summary of topological matter – the importance of holes, appreciating the absence of something as much as the presence.* In his textbook on condensed matter physics he provides a translation into modern physics of the opening lines of the *Daodejing*, the founding work of Daoist philosophy compiled around 400 BCE:

> *The physical theory that can be formulated cannot be the final ultimate theory.*
> *The classification that can be implemented cannot classify everything.*
> *The unformulatable ultimate theory does exist and governs the creation of the universe.*
> *The formulated theories describe the matter we see every day.*

Wen trained as a string theorist before switching paths to condensed matter physics. In 1990 he devised the concept of 'topological order'. It ties together knots, topological quantum computers and the emergent quasiparticles in the fractional quantum Hall effect.

Cutting the gordian knot

The children in my hometown of Ottery St Mary in Devon practise a number of strange rituals. On 5 November the townspeople, including children as young as seven, run through dense crowds with burning barrels of tar on their backs. In June we celebrate 'Pixie Day': children dress up as pixies, tie up the church bell-ringers, and drag them down

---

* Incidentally, Wen went on to offer me some invaluable Daoist career advice: he said that you shouldn't try to ride the waves of popular trends, because by the time you realise it's a wave the peak has already passed. Instead you should do what interests you and hope it creates waves for others.

the hill to a mock-up of a cave outside the town where pixies are said to originate. And on 1 May the children engage in a number of *Wicker Man*-style activities, including maypole dancing. It is this ritual that is of relevance here. A maypole is a tall wooden post with many ribbons fixed at its top. Children take hold of one ribbon each and dance around one another and the pole, braiding the ribbons. A simple dance has half the children dance one way round the pole, and half the other, weaving in and out and giving the ribbons a criss-crossed pattern.

The quasiparticles in the fractional quantum Hall effect can be thought of as dancing with ribbons attached to them. Imagine a magnetic field as woven from lines of magnetic flux. That's often how the idea of magnetic fields is introduced when first taught: placing a bar magnet on a table, we scatter iron filings around it and find that they fall in an intricate pattern, aligning with the lines of magnetic flux. This is also how Faraday first conceptualised magnetic fields; the archives of the Royal Institution in London contain his original workbooks in which he has scattered iron filings onto the pages and stuck them down with glue to show the lines of flux around a magnet. To understand how the lines of flux behave in the fractional quantum Hall effect let's imagine that the magnet, pointed towards the thin sheet of material, is a lot bigger than the sheet. In that case, the lines of flux can be thought of as all running parallel to one another as they pass through the sheet, like the bamboo forest setting of many a martial arts epic such as *House of Flying Daggers* (Figure 33 overleaf).

In those films people often leap gracefully between bamboo stalks that bend somewhat, without breaking; the protagonists' movements are masterful to the point of wizardry. Lines of magnetic flux are similarly able to bend and stretch, but not break.* In the fractional quantum Hall effect the electrons play the role of the

---

* If they broke they would terminate in magnetic monopoles – and we know those don't seem to exist.

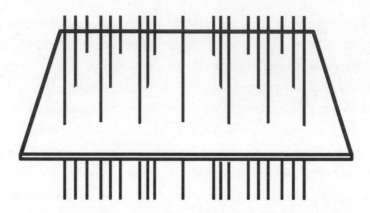

33: Lines of magnetic flux passing through a material like a
bamboo forest.

heroes jumping between bamboo stalks. But their movements are precisely coordinated in such a way that a set number of electrons must appear bound to a set number of flux lines (Figure 34). The number of flux lines and electrons is the same in every bundle, and these numbers dictate the measured voltage. For example, we might imagine that a single emergent quasiparticle takes the form of three electrons bound to two flux lines.

In general, lines of magnetic flux in a drawing do not convey any set amount of flux; all you can really say is that where the lines are more densely packed the magnetic field is stronger. But in the case of the drawings here, each flux line should be thought of as carrying a precise amount of magnetic flux. This amount, called the magnetic flux quantum, is a universal constant of nature. Just as all the quasiparticles must contain an integer number of electrons, they must also contain an integer number of units of this basic amount of flux. This model of 'composite fermions' is not the only way to think of what happens in the fractional quantum Hall effect, but it is perhaps the most intuitive to picture.

As the particles dance around one another their ribbons – the

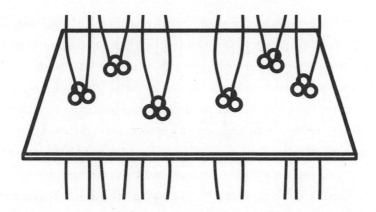

34: Trios of electrons bound together with pairs of flux lines.

lines of flux – entwine, and the pattern of entwined ribbons grants the set of particles a memory of where they have been. The past is encoded in the sequence of braids, just as the ribbons of the may-pole dancers tell the paths they have danced. It is this memory that allows the quasiparticles to function as computational devices. And their memory is resilient: it has a spell of protection cast upon it by topology for, however the particles might jiggle around, provided their paths do not unwind their memory remains intact.

Particles having a collective memory is pretty magical, and like all good acts of wizardry it also reminds us that rules are made to be broken (the Rule of Rebellion). To see how, recall that every particle is either a fermion or a boson. One way to understand the difference between them is how they behave when two identical particles swap places. This swap could be for any reason, but for the sake of argu-ment we can imagine grabbing hold of the particles and swapping them by hand. Now, swap two identical bosons and the quantum wavefunction describing them remains unchanged. This fits the clas-sical intuition that if you take two identical objects and exchange their positions the result is indistinguishable from the original. Swap two identical fermions, though, and their wavefunction changes.

Well, that's a bit strange, but it's not unheard of: if two identical owls face one another they could swap places and end up facing away from one another. What's clear, though, is that if you swap any two things *twice* you must get back to the original situation: swapping two things twice in the same direction is like circling one around the other. When the circle is complete, all is as it was (Figure 35).

But think of two maypole dancers. If one dancer circles another they return to their starting positions, but their ribbons are now entwined. Through this magical act the emergent quasiparticles in the fractional quantum Hall effect break the rule that all particles must be bosons or fermions. They are something entirely new: *anyons*.

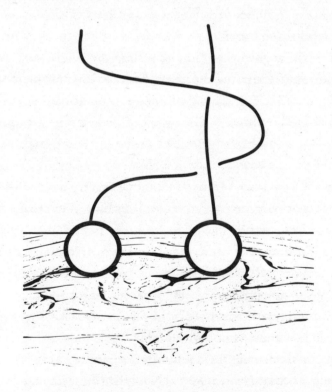

35: Two swaps should be the same as passing one object around the other.

The name is a pun: swapping two bosons is like turning a clock hand through a full turn – it looks the same. Swapping two fermions is like turning a clock hand through half a turn – it looks different, but doing it a second time returns the original. Swapping two anyons can be like turning the clock hand through *any* angle, hence anyon.

Different strengths of magnetic field in the fractional quantum Hall effect lead to different types of anyon. They require different numbers of swaps to return to themselves. It's hard to overstate how bizarre this is: take two objects, and pass one around the other – both are now different objects. It would be a phenomenal magic trick if it were performed by a magician, but it's performed instead by the universe. And, like all the universe's best tricks, this one is of practical use: it is this trick which may allow the creation of a scalable quantum computer.

Here's how it works. Tune the magnetic field in the fractional quantum Hall effect to give the right type of anyons; conjure an anyon and its antiparticle out of the quantum vacuum (a trivial task for the modern wizard); being antiparticles, if brought together the pair will annihilate, returning to the void. But instead conjure a second pair, and loop an anyon from the first pair around one from the second: you may find that neither pair can now annihilate. The reason is that they are no longer antiparticles to one another – both are now different objects.

The question of whether two anyons can annihilate is the evaluation of a simple piece of quantum logic, which is to say, the basis of quantum computing. By braiding many anyons around one another in prescribed sequences, arbitrarily complex programs can be encoded. It is the technology of the future encoded in that of the ancient past – the tying of knots. But if magical fiction has taught us anything, it is that using magic to bend the rules of the world usually leads to unexpected consequences: this is no exception.

A *matter of entanglement*

Changing the amount of magnetism in the fractional quantum Hall effect causes changes we can measure in our middle realm: for example, the voltage we would measure across the material. The electrons and magnetic field combine to give a fundamentally new emergent behaviour, without a precedent in the classical world. We have come a long way from earth, air, fire and water. In fact, the fractional quantum Hall effect contains within it an *infinite* number of distinct states of matter: a different behaviour for each of the different fractions, each composed of its own type of emergent quasiparticles.

In the bulk of the material a fractional quantum Hall state is an electrical and thermal insulator, like rubber or air. But the surface of the material conducts electricity, and this is true wherever the surface is. If you chip a bit off, or cut the material in half, or stick two lumps together, you change the surface, but the new surface is still conducting and the new bulk is still insulating. Imagine you had an orange that, no matter how you sliced it, always had a thick peel on its surface and segments inside. That would be magic, the sort of thing the Devil might give you if you sold your soul for a nice juicy orange.

If that is not remarkable enough, consider the phase transitions between different fractional quantum Hall states. The canonical story of phase transitions was presented in Chapter III. There we saw that matter is a state of broken symmetry: a crystal grows by breaking the continuous translational symmetry of a liquid, and phase transitions occur when symmetries change. But the set of possible phases in the fractional quantum Hall effect all have exactly the same symmetry, despite having measurably different macroscopic properties. They are referred to as *topological* states of matter, and they are connected by topological phase transitions.

The order in crystals allows them to respond rigidly to change:

push one end and the whole crystal moves. The states of the fractional quantum Hall effect feature a more subtle kind of order: namely, topological order. It is no less physical than the order in a crystal. Fractional quantum Hall states respond rigidly to change: they resist being compressed, just as a crystal does, but do so *without* spontaneous symmetry breaking.

When Xiao-Gang Wen introduced the idea of topological order he boiled it down to its essence, identifying its defining property. His characterisation was this: topologically ordered matter is defined by having long-range entanglement between its emergent quasiparticles. While all matter is entangled to some degree, topologically ordered matter is entangled in a practical way.

When I read about entanglement between pairs of elementary particles in *The Fabric of the Cosmos* all those years ago, I imagined it must be a property forever consigned to the microscopic world. I never dreamt that one day we would find it manifested in our middle realm, in lumps of stuff you could hold in your hand (albeit with some very well insulated gloves). Imagine: entire states of matter defined through entanglement. In one sense the connection between entanglement and topology is natural: when braiding one anyon around another, it doesn't matter how close they pass; all that matters is that the path closes to a loop. When entanglement affects the revelation of marbles in towers, it doesn't matter how close the towers are: the connections, once established, persist beyond space and time.

The onward paths

Topological matter is now a huge field of study in its own right, lying on the cutting edge of condensed matter research. The possibility of creating scalable quantum computers is but one motivation for it. Despite first being hypothesised in 1977, the first unimpeachable experimental observation of anyons was made only while I was writing this book. The experiments, conducted in 2020, were the first to directly observe anyons' remarkable defining property, their transformation upon braiding.[22] On the other hand, those paths that lead us to scalable quantum computation may not involve topology at all. Many other routes are being explored, including both engineering and theoretical approaches. At the time of writing, the leading contender is another route devised in part by Alexei Kitaev called 'magic state distillation'.*Many paths to quantum computing may lie parallel, their different approaches complementing one another. Whatever the route, my guess is it will not be long before we get there. Small quantum computers, with a few qubits, are already freely available to use online. You can literally create your own quantum superpositions and entanglements: you can write and execute your own quantum algorithms from the comfort of your home, by communicating over the internet with experimental labs around the world. The power of the microscopic world has already been brought into the middle realm.

Closing the loop to our opening question, whence do quantum computers derive their power? Deutsch's multiverse is one interpretation, but there are many more. What's agreed upon is that the power comes from wherever quantum mechanics derives its power beyond the classical world: wherever it is that the moon goes when nobody looks.

At present, quantum mechanics is heralding the end of Moore's law and is curtailing the growth of classical computers. In the near future we will learn to work *with* quantum mechanics to realise power beyond classical limits. But this is still a question of

---

* While not the only technical use of the word 'magic' in theoretical physics, this use has led to appealing article summaries such as 'There also remain some unanswered questions regarding the power of states with vanishingly small amounts of magic' (Stephen D. Bartlett, 'Powered by magic', *Nature*, vol. 510, pp. 345–7 (2014)).

maintaining perpetual growth – growth that is unsustainable in a universe of finite resources. If we are to survive into the *far* future, it must instead be balance we seek.

VIII

In Search of the Philosophers' Stone

Veryan sailed into a large mercantile port on the coast. She arrived on a crisp autumn morning just as the sun began to burn off the mist. The buildings, perched firmly on ancient tarred timbers, were constructed from pale yellow stone. Their tile rooves were of a distinctive red, a counterpoint to the turquoise water. There was a bustle of morning traffic along the city's canals as merchants received crates of tea and spice, and fishing boats brought in the morning's catch. Veryan furled the jib before smoothly bringing her boat to rest. As she secured it to its mooring she saw the familiar shadow of her friend Beatrice waiting for her on the dock.

The tangle of cobbled streets was dense with people going about their tasks before the burning heat of the midday sun. Despite the crowds, Beatrice could walk for hours through her city without meeting a soul. She and Veryan passed effortlessly through the city's narrow backstreets, arms linked in the local custom. The high yellow walls formed a vast maze, the occasional window or doorway serving as a reminder that the narrow streets were lined with homes. The comforting familiarity could lull visitors into a dreamlike state. Entranced, some sleepwalked into blissful non-return, led by the playful spirit of the city which skipped joyfully ahead of them. Veryan knew that she would need to give herself over to it if she was to find what she sought. Around her neck hung a heavy talisman. Grasping it, the weight would ground her when she needed to return.

Having rested out the heat of the early afternoon in a patch of shade, Veryan and Beatrice continued on. It was early evening when Veryan gradually became aware that the wall to her left had been curving gently away from her for as long as she could recall. It was a closed loop a mile or so in length at the heart of the city. This was the entrance to the catacombs, whose recesses were the caverns that granted access to the great library. Grasping her talisman, she grounded her thoughts. Along this wall there would be a door. While in plain sight, only those with knowledge of the word could see it. But Veryan had this knowledge — the word had been told to her by the plants and creatures of the forest an age ago. And so it was that Veryan crossed the threshold, Beatrice skipping off to greet many other friends who would be visiting her that evening, some of whom would be staying for a very long time.

∼

The inevitability of loss

Why don't we just cover the Sahara in solar panels? Well, it's a valuable and unique ecosystem. But why don't we generate energy renewably by taking advantage of local conditions, before sending it all over the world? Solar power in the Sahara, geothermal in Iceland, wind in Chicago (OK, that last one's a myth). The short answer is the second law of thermodynamics: moving energy around necessarily involves losing a lot of it on the way, mainly as heat, but also as sound and vi-brations. You can hear power lines hum, for example: if you stick one end of a strip light in the ground below a power line it will light up. In the 2006 film *The Prestige* a similar feat performed by the inventor Nikola Tesla leads him to be declared a *true* wizard – a person who can actually do the things a magician pretends to. The effect is due to the 'corona discharge' in which the lines dissipate power through the air. In 2021 the United States alone lost around $31 billion dol-lars' worth of electrical energy in the process of transmission and distribution: enough to run every streetlight in New York for a mil-lennium.[23] While a huge amount of effort goes into minimising loss, the second law seems to tell us it is inevitable. Throughout living memory, technology has advanced at an exponential rate. This has coincided with the exponential growth of the human population, with both beginning around the time of the Industrial Revolution. But the UN estimates that the world's population will stabilise by 2100. The previous two chapters gave instances of how condensed matter physics might permit humanity's technological advancement to continue unabated. Yet as we look to the far future, it seems rea-sonable to hope that a stable population might instead seek technol-ogies to reach equilibrium with their environment.

In practical terms this requires us to pass energy from place to place without loss. It might be seen as a continuation of the ultimate quest of the alchemists: the transmutation of lead into precious metal, to be achieved by finding the philosophers' stone. Physical transmutation was just one aspect of the stone's power, however. Its higher function was the granting of immortality – freedom from loss. Similar searches appear from ancient Sumeria to the Magi of Persia, Daoists in China, alchemists in Europe, and the Dogon of Africa. In all cases the belief is associated with the seemingly magi-cal powers of metalworkers; this connection was put eloquently by Mircea Eliade in *The Forge and the Crucible*, a history of alchemy:

It is significant that the mastery of fire asserts itself both in the cultural progress which is an offshoot of metallurgy, and in the psycho-physiological techniques which are the basis of the most ancient magics and known shamanic mystiques.

In 1911 Dutch physicist Kamerlingh Onnes realised the alchemists' dream, transmuting base substance into precious metal. His substances were mercury and lead – both staples of alchemy. His success lay not in the application of heat, but its utter banishment, cooling them to the coldest temperatures ever achieved on Earth. Rather than gold, Onnes's matter transformed into something far more valuable: a superconductor.

Superconductors possess the ability to convey electrical currents entirely without loss. Like unsuspecting visitors to the port in the opening passage of this chapter, an electrical current established in a closed loop of superconductor will continue forever without loss. Onnes's superconductors were probably the first ever to exist on the Earth, and possibly the first to exist in the universe outside neutron stars. Learning how superconductors cast their spell has been one of the greatest success stories in physics: a bizarre and unfathomable set of phenomena explained in detail with a simple theoretical model. Yet the story of superconductors is far from finished: the simple theoretical model tells us that they can only ever be found at extremely cold temperatures, not to be found on Earth even in the coldest desert nights. Understanding how to overcome this impasse is another spell that condensed matter physicists are still learning. It is the search for the modern philosophers' stone: a *room temperature* superconductor.

This final leg of our journey will take us into the far future of condensed matter physics, and there are no guarantees of what we will find. But we can begin on well-trodden ground, by looking at a type of matter which has fascinated me since I heard tell of its magic as a young child: the superfluid.

Secret preparations and superfluids

I remember first hearing about superfluids when I was in primary school, aged about eight. They sounded like the most magical things ever. While I already knew I wanted to be a physicist, with hindsight I wonder if hearing of superfluids set me on the path to condensed matter physics. Specifically I remember hearing that if you put a superfluid in a bottle, it runs up the walls, out the top and down the outside. The superfluid can leak through the sides of the bottle, even if the bottle could contain any normal liquid. And I remember hearing that superfluids maintain their fluidity even at absolute zero. I forget how I heard these facts in mid-90s rural East Devon, but I'm glad I did. And fortunately, unlike a lot of what passed for facts among the eight-year-olds of Ottery St Mary County Primary School, these turned out to be true.* Helium is the only substance known to become a superfluid through cooling alone. Other super-fluids have been made, but they require exotic set-ups or phenomenal pressures: the neutrons in neutron stars are believed to form superfluids, for example. Nevertheless, these exotic materials are beginning to find practical uses on Earth. One quantum computing start-up even plans to build functional qubits employing superfluid helium.

The magical powers of superfluids stem from their lack of viscosity. Viscosity is a measure of how runny a fluid is: Marmite is viscous, for example, whereas water is not. Quicksand, threatening to drown unsuspecting desert adventurers until its mysterious disappearance from popular culture in the mid-1990s, derived its power to terrify from its variable viscosity: a mix of sand and water, it is

---

* On the other hand, a story that a giant tarantula had escaped from London Zoo by picking the lock of its cage using its 'sting', and was last seen on the London-to-Exeter train, proved a fallacious rumour.

viscous to the point of near-solidity until stood on, at which point it flows freely, pulling its victim to their doom.

Superfluids have exactly zero viscosity. Their ability to overflow their container is called the fountain effect; normal liquids actually show a similar effect, though to a much smaller extent: owing to surface tension, liquids naturally flow up the walls of their container very slightly, which produces the meniscus, the upward curve at the edges of the tea in a cup, for example. Fortunately for tea drinkers, the viscosity of normal liquids stops them making it too far up the walls. Nevertheless, this classical effect can be put to use as a practical bit of magic.

To enact it you will need to return to your local tavern. You are certainly barred at this point, so you'll need to go in disguise, perhaps as an elderly bookseller – unless, of course, you *are* an elderly bookseller, in which case you could perhaps go as a desert explorer recently returned from their travels imbued with a 'funny sense of fun'. Find a way to work a cork into your disguise. Engaging an unsuspecting tavern-dweller in conversation, expertly direct your conversation to the topic of corks. For example, you might tell them that, on your travels through the desert, you chanced upon a magician: this magician revealed to you that it is ordinarily impossible to balance a cork in the centre of the surface of a drink, but they had learnt a spell by which the cork might be convinced to hold to the centre. Your trick will work with almost any drink, but let's suppose your companion has a tankard of mead. Give them your cork, and let them try to steady it in the centre of the surface of their mead. They will indeed find it impossible: the cork floats to the highest point on the surface, which, owing to the meniscus, is at the edge. Now you reveal the spell taught to you by the magician. The trick is to overfill the tankard slightly, so that the meniscus inverts. The mead rises above the rim of the tankard, with its highest point in the centre, and the cork will now naturally move to the centre of the surface. If you favour the dark arts you can readily work this into a profit-making wager. In general, a meniscus forms because the molecules in the fluid are more strongly attracted to the walls of the container than they are to one another. By climbing the wall of the container they can increase their contact with it. Water does this in glass, as do many water-based liquids such as milk and honey (hence mead, a honey-based alcoholic beverage, should work). Some substances have a stronger attraction between their molecules than to their container; in this case the meniscus is higher in the centre than at the edges, when the container is not overfilled. Mercury can be seen to do this in a thermometer.*

The ability of superfluids to climb walls is an extreme version of the meniscus effect. Their other abilities are harder to understand, and are inherently quantum in nature. In Chapter V, I attributed helium's lack of solidity at absolute zero to quantum fluctuations. A more precise statement depends on the particular isotope: helium-4 or helium-3. Different isotopes of an element have the same number of protons but a different number of neutrons. Helium-4 has two of each; below 4.2 K it becomes a new state of matter, entered by a phase transition called Bose–Einstein condensation, transforming into its alter ego, the superfluid.

Recall those tales of the Fermi sea. The Pauli exclusion principle states that no two identical fermions can occupy the same quantum state. This means that when many fermions get together, some end up with large energies even in their lowest-energy state, like lazy wizards trekking up to high rooms in a tower when the lower ones are occupied. But helium-4 atoms do not behave like fermions: they behave like bosons.

---

* I stand by my statement that the cork effect should work with almost any drink: since mercury is toxic it would not generally be drunk. This said, China's *Orthodox Histories*, compiled over two millennia, document at least ten emperors who either died or were driven insane through drinking mercury-based compounds presented to them as elixirs of life, in search of immortality.

Bosons are named after Satyendra Nath Bose (1894–1974), a Bengali physicist who founded the field of 'quantum statistical mechanics'. Despite seven Nobel Prizes so far being awarded for work directly facilitated by his 1924 discovery, Bose himself never received this recognition; it would be another six years before the Nobel committee first awarded the physics prize outside Europe or the United States, and a quarter of a century before they did so for a second time.*

Bosons are not subject to the Pauli exclusion principle. They are perfectly happy existing in the same quantum state as one another, and so at low temperatures they all choose the lowest-energy state. Bosonic wizards would all pile into the lowest room in the tower. If bosons additionally feel an attraction to one another, as helium-4 atoms do, they actively *encourage* one another to join them in the lowest-energy state. The resulting state of matter is called a Bose–Einstein condensate, and superfluid helium-4 is an example. The bizarre macroscopic properties of superfluids are inherently quantum, but you can nevertheless see them on a large scale in our middle realm. Isn't that incredible? How can it be?

Recall that all the information about a quantum particle is contained in its wavefunction. Each of the particles in a Bose–Einstein condensate picks the same wavefunction; in helium-4 these particles are individual helium atoms, and the wavefunction they choose corresponds to the lowest possible energy. Since all the particles have the same wavefunction, it makes sense to describe the *entire condensate* with something resembling that same single quantum wavefunction. This is the essence of the magic of superfluids – it is how quantum phenomena manifest on everyday scales.

---

* Initially failing to have his results published, he wrote to Einstein, who immediately understood their importance. Einstein personally translated them from English to German and had them published in Bose's name in a renowned journal, following up with his own work. The collective behaviour of many bosons is now said to be described by Bose–Einstein statistics.

But when we turn to helium-3 there is already a mystery, because helium-3 atoms behave like *fermions* rather than bosons, yet at really, *really* low temperatures, about 0.0025 K, helium-3 *also* becomes a superfluid. This doesn't make any sense: fermions behave like the wizards in the tower; what are they all doing on the ground floor? They must have found a way to behave like bosons. Understanding how is the key to superconductivity, where it is electrons (fermions) rather than helium atoms that are important.

Superconductors

Some important subtleties aside, superconductors are superfluids with an electric charge. While superfluids are exceedingly rare, superconductors are rather common, albeit at very low temperatures. In fact, all metals are expected to superconduct at low enough temperatures unless there's some special reason for them not to, in much the same way that all liquids are expected to freeze to solids when cold enough. If we lived at temperatures close to absolute zero, superconductivity would be routine; familiar; *boring*. Ours would already be a world without loss – at least in the sense of electrical currents. But the Sahara is far from absolute zero.

Cool a lump of metal sufficiently and the electrons dramatically change their collective behaviour. The positive ions remain a crystal, so the material overall maintains its shape. Yet the new behaviour leads to a host of magical phenomena, all of which have been given appropriately sci-fi names.

First, supercurrents: superconductors have zero resistance to the flow of electrical current. Not nearly zero, or zero to experimental precision; exactly zero. If you set a supercurrent flowing around a ring, you could return at the end of a long and well-lived life to find it going exactly as you left it. Importantly for us, supercurrents could also be carried arbitrarily far along superconducting power lines without loss.

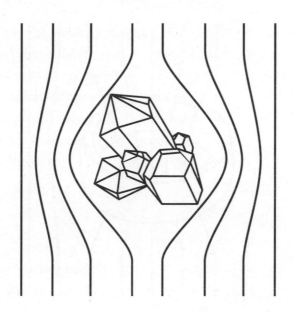

36: The Meissner effect: magnetic flux lines are banished
from a superconductor.

Second, the Meissner effect (Figure 36). Apply a magnetic field
to a lump of lead, and the field passes through it like normal. But
cool the lead all the way to 7.2 K, and magic takes hold: the lead
transmutes into a superconductor, banishing the magnetic field
from within itself. This is the Meissner effect: magnetic fields cannot
exist inside superconductors. Instead, the lines of magnetic flux are
pushed around the outside.

Third, flux trapping. Take your lump of lead and forge it into a
ring (Figure 37 on the following page). Now take that ring and place
it in a magnetic field, allowing the lines of magnetic flux to thread
through it. Cool the ring as before, and allow it to transform into its
super alter ego. The superconductor banishes the magnetic field as
before – but this time, some of the field is banished to the *inside* of
the ring. It becomes trapped inside and cannot be removed, as to do
so would require the flux lines to pass through the superconductor

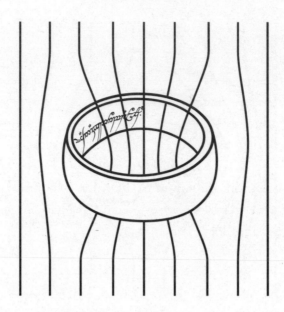

37: Flux trapping: lines of magnetic flux become trapped in a superconducting ring.

itself, which is impossible. This is flux trapping. The ring is held in place by the magnetic field: turn off the field, and the field lines passing through the ring will remain, threading the loop.

Fourth, flux quantisation: measure the magnetic field that was trapped in the ring, and you will find it is precisely quantised, meaning it appears only in integer multiples of some smallest amount: a universal constant of nature called the 'magnetic flux quantum' (possibly the most science-fiction name in science fact). The magnetic flux passing through the ring can be one flux quantum, or two, or three, but not 1.2, say.

There is an important distinction between two types of superconductor. Type-I superconductors don't allow a magnetic field within them under any circumstances: if the magnetic field is made stronger and stronger, eventually the superconductivity just gives up and the material abruptly turns back into a normal metal. Type-II superconductors similarly resist smaller magnetic fields, but rather than completely give up at larger fields they begin letting them in as bundles of the magnetic flux quantum. This does not contradict the phenomena we just encountered: wherever the flux passes through the superconductor, that region returns to being a normal metal. You can think of it as a lump of superconductor turning into a ring of superconductor with the flux threaded through the hole.

Superconductors might sound like they belong in the engines of fictional time machines but they actually have a huge range of practical uses. They have something of a love–hate relationship with magnets. Spinning a superconductor generates a magnetic field perfectly aligned to the axis of rotation; this can function as an incredibly sensitive gyroscope – used, for example, in Gravity Probe B, a satellite-based experimental test of general relativity. Recall that an electric current flowing in a loop generates a magnetic field; since supercurrents flow without loss, they can generate huge magnetic fields. These superconducting magnets find many uses: they are, for example, present in every MRI scanner in the world. They are also used in NMR experiments to control and measure the magnetic fields of the nuclei in atoms. Superconducting magnets are also used to control the plasma within nuclear fusion reactors, and to accelerate elementary particles in the Large Hadron Collider to phenomenal energies not seen since the early universe.

An undeniably magical use of superconductors is magnetic levitation. The idea is an ancient one: Pliny the Elder describes an attempt to build a levitating statue using lodestones, which also power the flying island of Laputa in Jonathan Swift's *Gulliver's Travels*. Neither could have worked because Earnshaw's theorem tells us ferromagnets cannot cause static levitation. And yet, recall the floating crystal I saw in the office in St Andrews: it worked because it was a diamagnet, something that becomes magnetic so as to oppose any applied magnetic field. It was a crystal of pyrolytic graphite, the strongest-known diamagnet under ambient conditions. But a superconductor *perfectly* expels a magnetic field – this is the Meissner effect. Superconductors are therefore *perfect* diamagnets. Every superconductor is over 2,000 times more diamagnetic than pyrolytic graphite.

Superconductors are already being put to practical use for magnetic levitation, commonly referred to as maglev. There exist a number of high-speed maglev trains: the fastest train in the world is the Lo series superconducting maglev in Japan. It travels at 375 mph (603 km/h) and requires only one-eighth the acceleration distance of conventional trains. These feats are possible due to the lack of energy that is lost to friction as the train flies above the track. The major remaining source of loss is air resistance; plans exist to set maglev trains in vacuum tunnels, allowing further massive speed increases. Commercial levitation does not use the Meissner effect; rather, superconducting magnets generate large magnetic fields which induce paramagnetic fields in the tracks. Earnshaw's theorem is circumvented once again, this time because the train is moving.

Superconductors certainly meet the practical standards of magic required by wizards. The explanation of how they arise is more magical still.

La Danse Macabre

Superfluid helium-4 forms through Bose–Einstein condensation. This only works for bosons; electrons are fermions, so they must find their own route to superfluidity. They do so through a curious and coordinated dance, which has no familiar analogue. To gain some understanding of it, we can turn instead to the supernatural, to dancers who go by many names: The Good People, The Fey, Tylwyth Teg. It is the dance of the fairies (Figure 38).

Stepping into a fairy ring can be dangerous. Modern science

understands these rings of mushrooms to be the surface growth of a larger underground fungus – but ancient folklore has it that they are the dancing places of fairies. If you are foolish enough to step into the ring you will hear their enchanting music; you will likely find yourself forced to dance forever. The dance of the fairies is referenced in *A Midsummer Night's Dream*, in a wedding attended by both humans and fey alike. The scene, captured in William Blake's 1785 painting *Oberon, Titania and Puck with Fairies Dancing* (these being the names of some of the principal fey), was based on accepted ideas: Welsh tradition has it that when fairies are found in a group, they are inevitably dancing; while in his 1828 book *The Fairy Mythology* Thomas Keightley describes a dance 'to that ravishing fairy-air which charms the mind into such sweet confusion'.

Similar rings appear in the world's oldest desert, the Namib; closely packed circles of barren sand are surrounded by tall grass. Local legends attribute them variously to the dancing of fairies or the footsteps of gods. Their scientific explanation is disputed, but the leading theories are fascinating. A 2017 study by researchers from Princeton modelled the fairy rings' growth as a combination

38: Olaus Magnus's 1555 *On Nocturnal Dance of the Faries, in Other Words Ghosts.*

of competition between termite populations and the coordinated growth of vegetation.[24] The rings pack into a pattern where each has on average six neighbours, resembling a honeycomb, the densest possible packing of circular disks in the plane. It is a beautiful example of emergence: each termite and plant carries out simple tasks, but the large-scale result is a near-optimal packing of fairy rings.

Let us imagine a foolhardy desert caravan has strayed into a fairy ring. Hearing the sweet music the caravan's members join the eternal dance. Everyone, fairy and human, dances swiftly; all possible directions of dance are represented by the humans, and separately by the fey, and all feel the collective stamping through the desert sand. The intoxicating vibrations cause an attraction between humans and fairies alike; yet the miracle of the enchantment is that each of them finds themself bound to only one partner. Each person has a fairy partner and vice versa, and partnerships are formed between whoever happens to be dancing in the opposite direction in that moment. Partners may well be far apart in the crowd, but their eyes meet, and their fates are bound; yet the moment is fleeting, for all are jostling and merrily changing direction. As soon as any human changes direction, they immediately find a new partner in whichever fairy is dancing in the opposite direction, and whenever any fairy changes direction they find themselves similarly re-paired.

Allow me to explain the analogy. In a normal metal only those electrons near the top of the Fermi sea are involved in conducting electricity. These electrons have the most energy and are the fastest moving. Typically their speeds are all about the same, but they travel in different directions. When the metal is cold enough it is these electrons which begin their superconducting dance. Both humans and fairies represent electrons − negatively charged particles that also have a magnetic field (a quantum spin). Measure the spin along a chosen direction and you will always find it either pointing along that direction, or opposite to it; humans and fairies represent these two directions of spin, respectively.

Humans and fairies are attracted through vibrations stamped into the desert floor; similarly, the electrons feel one another's vibrations passed through the crystal lattice. Being of like charge, electrons would naturally repel, but these good vibrations lead them instead to *attract* one another. This attraction is the first remarkable feature of the theory of superconductivity. It is attraction in the familiar sense in physics: a force pulling the particles towards one another. But its appearance here is bizarre, because two elementary electrons would feel a force pushing them apart, and it is not obvious what counteracts this.

A story that is told to explain the attraction goes as follows. Recall that a metal is a mix of earth and fire: it has a periodic crystal lattice occupied by positive ions, surrounded by a sea of negative electrons to which they are attracted. In a superconductor, so the story goes, as an electron travels it pulls the positive ions slightly closer towards it. As the electron continues on its way it leaves behind a slight concentration of positive charge because the heavier ions take a while to relax back to their original positions. This region of positive charge then attracts a second electron. So by causing a vibration in the crystal lattice, the first electron effectively attracts another. While this basic picture of superconductors is certainly intuitive, it shouldn't be taken too literally because there are various other aspects of the phenomenon it doesn't account for. Really superconductivity is a quantum phenomenon, and classical analogies can never quite capture it.

The end result of these attractions is a second remarkable feature: the electrons bind together into pairs. It is intuitive that two things which are attracted might want to bind together. But the strange thing is that many, many electrons are attracted to one another, yet they still bind into pairs, rather than trios, quartets or a collective lump. Each electron partners with the electron with opposite spin travelling in the opposite direction to itself, just as each human partners with the fairy dancing in the opposite direction. But it is not individual electrons that pair up: a person does not stay partnered to an individual fairy – rather, they are always partnered with whichever fairy dances in the opposite direction, but the directions of individuals are constantly changing. This distinction is important. If it were individual electrons which paired, the pairs could be disrupted by disorder and impurities, much like electrons in normal metals. But because it is directions which pair, jostled electrons can drop in and out of partnerships. If an electron changes direction it finds itself with a new partner with opposite direction: as the humans and fairies jostle in their complex dance, pairs are constantly changing, but all are always paired.

For this to work there must be a huge amount of coordination between the dancers, but it comes about spontaneously. For instance, partnerships must scatter in opposite directions at the same time, and must find new partnerships available at that same instant. Einstein told us that things cannot really happen instantly at a distance, as nothing can travel faster than light, but that makes it all the more remarkable that the dance can continue to be so perfectly coordinated.*

A pair of dancers need not be nearby: paired electrons in a superconductor can be very far apart – typically, thousands of atoms away. In fact, the separation of partners within a pair is much larger than the spacing between different pairs, just as two dancers can be partnered across a fairy ring full of other dancers. As a result, many pairs of electrons can happily coexist in the same place. Thinking classically, most of the 'pair' is empty space.

---

* I was once speaking online with a large group of friends when we attempted to sing 'Happy Birthday' to someone: the tiny delays between each of us led the song to slow to a halt.

The steps of this dance were first understood by John Bardeen, Leon Cooper and J. Robert Schrieffer in 1957; the explanation is known as the BCS theory of superconductivity, after their initials, and is one of the great success stories of theoretical physics. It is elegantly simple, both in terms of its mathematics and the surrounding physical story. BCS theory is a complete quantum description of the behaviours of huge numbers of interacting particles; it perfectly explained a wide range of experimental observations that had confounded physicists since Onnes's discovery of superconductivity four decades earlier; and it made a number of new predictions that were quickly tested, confirmed and put to practical use. Its creators were duly awarded the Nobel Prize in 1972.

The formation of attracting pairs is key to superconductivity. These pairs really are bound together through their attraction to one another, a bit like how the Earth and the Moon are bound by their gravitational attraction, or the way in which two dancers can swing in circles while holding hands. The difference is that in those cases the members of the pair stay a fixed spatial distance from one another, while the pairs in a superconductor are linked instead by their relative directions of travel.* The stage was set for BCS theory when Leon Cooper calculated that any attractive interaction between many electrons would lead them all to bind into pairs. These are now known as Cooper pairs: cooling the metal, there is perfect conduction as soon as the first pairs exist.

Cooper pairs

Cooper pairs are what allow electrons – fermions – to form superfluids as if they were bosons. In Chapter VII we saw that a pair of

---

* By the terminology of Chapter III, the pairing is in reciprocal space rather than real space.

identical bosons looks the same when they swap, while a pair of identical fermions looks different. In fact, the wavefunction describing them picks up a minus sign. But minus times minus is plus – so *pairs* of identical fermions can resemble bosons. This grants Cooper pairs boson-like properties: importantly, they are all able to occupy the same lowest-energy state. The wizards in the tower have learnt the dance of the fairies: they find a way to partner that allows them all to pile into a room on the ground floor, and, in doing so, they lower their energy.

Cooper pairs helped me gain a deeper understanding of what it means for something to be a quasiparticle – because they are *not* quasiparticles. According to the definition of quasiparticles in Chapter I:

> *An emergent quasiparticle can exist by itself above the ground state of a material, and cannot be reduced to other things with that property.*

Well, Cooper pairs can't be broken apart into smaller things: despite being pairs, they only emerge through the interactions of many particles – just as the huge number of dancers in the fairy ring must all be dancing for the pairs to form. That's the magic of the theory. But they are not quasiparticles, because quasiparticles are excitations *above* the ground state, whereas Cooper pairs are *in* the lowest energy state. They are the superconducting equivalent of the still Fermi sea, rather than the excitations above it.

So what are the quasiparticles in superconductors? Cooper pairs *can* be broken apart, given enough energy. But when they break apart, the result is not two individual electrons. It is something much weirder; a pair of 'Bogoliubov quasiparticles', named after Nikolay Bogoliubov and sometimes called bogoliubons. A bogoliubon is a quantum superposition of an electron and a hole: an electron and its own absence. This leads to some bizarre properties. For example, the electron and hole have opposite electric charge; a bogoliubon, being a superposition of the two, does not have a well-defined charge. Experimentalists have found the charge to be anything between an electron's negative charge and a hole's positive charge.[25]

Supercurrents, carried by Cooper pairs, allow lossless transmission of power over arbitrary distances. But there is a problem. BCS theory predicts a highest possible temperature at which superconductors can exist. At about 40 K, this temperature is impractically cold: the coldest recorded temperature on the surface of the Earth is 183K, and that was in Antarctica. While lossless power lines could already be built from superconductors, as it stands they would use more energy to cool than would be saved by using them.

What seemed like a sure route through the dunes to a lossless future turned out to be quicksand. To get a clue how to proceed, we can look back over the various definitions of matter we have met so far on our journey.

A friend in many guises

Matter has appeared repeatedly in this book, veiled each time in a different guise. Superconductors don these many outfits simultaneously.

In Chapter II matter took the form of that which emerges from the interactions of huge numbers of particles. Superconductors definitely fit that definition, because it is only through the interactions between many electrons and phonons that they come about. In Chapter IV matter appeared as a balance between minimising energy and maximising disorder: low-energy stability perturbed by thermal fluctuations, while in Chapter V we saw that quantum fluctuations can perform a similar job, allowing helium to avoid solidity at absolute zero, going against the classical intuition that absolute zero is the temperature at which motion stops. In fact helium is not a liquid at absolute zero, but a superfluid. Superconductors, too, survive to absolute zero; far from motion stopping, it can become perpetual. But it is not perpetual motion in the forbidden sense: this 'flow' occurs in a strange, quantum sense we have met before – and in precisely the same sense in which an electron flows in its orbit around the nucleus of an atom.

In Chapter VI we met the spin ices: long-range correlations without long-range order; superconductors, as is more typical, have both. In Chapter VII we met topologically ordered matter, defined by long-range quantum entanglement. When topological order came to be understood it was suggested by Xiao-Gang Wen and others that it may have been seen nearly a century before – with Onnes's discovery of superconductivity. While not universally accepted, the essential idea, as proposed by Wen and elaborated upon by others, is this. Recall that type-II superconductors allow sufficiently strong magnetic fields to pass through them, but only in quantised amounts: the flux lines pass through in bundles of the magnetic flux quantum. Now recall the classic example of topology: a wizard blowing a smoke ring – the smoke follows an air vortex, which is like a tornado which has been bent into a loop so that the ends meet. Combine these ideas and imagine bending the bundle of magnetic flux into a loop which exists entirely inside the superconductor. You then have something resembling a smoke ring made of magnetic flux – and, like the smoke ring, the flux ring is stable, meaning it can bend and stretch without breaking open. In the fractional quantum Hall effect, passing one anyon around another can change both into something else. In much the same way, if a bogoliubon passes through the ring and returns to where it started, both the ring and the bogoliubon can change their form.

However, the canonical definition of matter, and the example to which we have returned throughout this book, is the example of a crystal growing from a liquid in Chapter III. This is an act of *spontaneous symmetry breaking*. The liquid has a continuous rotational symmetry: it looks the same from all angles. A crystal has only a discrete rotational symmetry: it only looks the same along certain special directions. When the crystal grows, the symmetries of the liquid break, like an egg rolling off its tip in a random direction. The crystal's atoms form spontaneously into a periodic arrangement. The result is long-range order: knowing the position of one atom is enough to know the positions of all others. This grants the crystal *rigidity*: push one end, and the entire crystal moves. An exact analogue occurs when a superconductor grows from a metal. Superconductivity develops through a process of symmetry breaking. But when a crystal grows it is clear which symmetries break, and the rigidity is intuitive: push one end, the whole crystal moves. So what symmetry breaks when a superconductor grows from a metal? What rigidity results?

Only a phase

To understand the symmetry breaking in a superconductor, it is necessary to go deeper into the quantum description of the microscopic world. Quantum mechanics describes things using wavefunctions. A classical ocean wave is also described by a wavefunction, which quantifies the height of the wave at each point on the water, and how far along each point is in its cycle of up and down. This second quantity is called the 'phase'. Just as the phases of the moon cycle periodically, so each point on the water's surface cycles from high to low and back again. The quantum wavefunction also has a phase, and it is vital to understanding symmetry breaking in superconductors.

Let us return to those foolish humans enticed into the fairy ring, dancing their perpetual dance. Each person and fairy has a pocket watch. When the dance begins all pairs synchronise their watches perfectly. Now, I know what you're thinking: on those rare occasions when people are reported to have escaped fairy rings, they return months or years later, after what seemed to them like only a few hours. Time is meaningless in this place; how can the pairs know the correct time, down to the second? The trick is that they don't need their watches to tell the *correct* time – they only need them to tell the *same* time. All that matters is that the pocket watches are synchronised between pairs. The pairs agree an arbitrary starting time on their watches down to the second – a position for the second hand – and set them going. The second hand of a watch cycles just like the phases of the moon; in this sense, the pairs randomly pick a phase (starting angle of the second hand), but if any pair is off they will find themselves out of time with the others, which in the coordinated mayhem of the fairy ring would not do. If a pair's watch began to lose time, the pair would quickly feel it as they fell behind all others, and they would be pulled back into step, adjusting their watches accordingly. In this sense all dancers collectively resist any change of phase, and so there is a rigidity to the collective dance: even though the choice of phase was arbitrary, once chosen it is very difficult to change. This is exactly analogous to how crystals' rigidity grants their ability to resist shear forces: superconductors' phase rigidity grants their supercurrents the ability to resist loss.

It is the familiar dance of spontaneous symmetry breaking: an egg balanced on its tip falls in one direction only, choosing one direction to the exclusion of others and breaking the rotational symmetry it had before it fell. Similarly, when a crystal grows from a liquid, the atoms choose particular locations to the exclusion of others. And when a superconductor grows from a metal, the Cooper pairs choose one phase for their quantum wavefunction to the exclusion of all other phases.

The fact that all Cooper pairs choose the same phase means that the entire superconductor can be described by this phase on everyday scales. This is what fascinated me as a child: superconductors and superfluids live in our middle realm – they're lumps of stuff you can hold in your hand, using your (now well-worn) insulated gloves. Yet this entire lump of stuff has the same quantum mechanical phase; it is quantum madness scaled up, through the magic of rigidity.

The collective wavefunction of superconductors and superfluids is called a 'coherent state'. As we saw in Chapter V, when a quantum system is coherent it is able to work its magic. There is a beautiful subtlety here, because the phase of a quantum wavefunction is meaningless by itself: it cannot be measured in any experiment. Only the relative phase between different wavefunctions has a meaning.

How does this fit with the idea that superconductivity results from choosing a phase? Here's what happens. When the superconductor forms, it picks a phase relative to every other superconductor in the universe. This may seem abstract, but it is directly measurable, and the measurement is one of the most practical effects in all of quantum mechanics.

Josephson junctions

There is a simple device that measures the difference in phase between two nearby superconductors. It is called a Josephson junction and it has many practical uses. Remember in Chapter VI when I said Ritika Dusad's measurements of magnetic monopoles were the most sensitive detection of magnetic flux ever performed? Well the device she built employed a superconducting quantum interference device, a SQUID. This is nothing more than two Josephson junctions connected in a ring. Josephson junctions are also used to make the most precise measurements of electric charge ever performed: the volt was formerly defined using a measurement taken on a Josephson junction. They will soon become the standard devices for capturing images in astronomy and astrophysics, and they are a leading candidate for practical qubits in quantum computers. When you picture a quantum computer, do you imagine a menacing golden squid hanging from a ceiling? If so, the quantum computer you're imagining is probably built using Josephson junctions.

So what are they?

Josephson junctions are very simple. They are just two super-conductors placed close together, separated by a narrow non-super-conducting gap. But what results is rather magical. And like all good magic, when it was discovered it was too much for most people to believe. The effect was predicted in 1962 by a twenty-two-year-old theorist named Brian Josephson. John Bardeen, later a double Nobel laureate and the B in BCS, publicly came out against Josephson's prediction, but within a year the prediction had been confirmed experimentally, and by 1973 Josephson had been awarded his own Nobel Prize. The effect is this: when you place two superconductors close to each other, an electric current spontaneously develops between them. Apply a voltage, and the current sloshes back and forth between the two superconductors at a frequency proportional to the voltage.

Recall those humans in the desert fairy ring, their dance co-ordinated to the second by their pocket watches. Now imagine there is a second fairy ring nearby, and the same dance is going on. Occasionally a pair crosses the gap, dancing from one ring to the other. Now, that's strange enough, because the dance shouldn't exist outside the ring. But it gets stranger. Even if both rings contain the same number of dancers, dancing the same dance, there will be a net flow from one of the rings to the other. How can that be, when the dances are identical? Isn't there a perfect symmetry between them?

Here's the trick. All dancers in each ring picked the same arbitrary starting time on their watches: the same phase. Once chosen, the phase is rigidly maintained by the dancers. However, being chosen randomly, the collective phase will likely be different between the two fairy rings. The watches of one ring's dancers will be synchronised earlier than the other, breaking the symmetry between the two rings.

Allow me to break down the analogy. Place two superconductors near one another: occasionally Cooper pairs will cross from one to the other. The reason Bardeen didn't believe this could happen is that the gap between them is not superconducting; Cooper pairs emerge from the collective behaviour of the superconductor – how can they exist outside it? But the Cooper pairs don't dance across the gap: they tunnel through it quantum mechanically. A supercurrent develops spontaneously between the two, its rate of flow set by the difference in phase of the superconductors. There is no classical analogue to this process. However, if you will permit me to let the analogy slip a little, I can give an idea how a phase difference might lead to a net flow.

Imagine that occasionally a pair of dancers will find themselves dancing outside their ring, but that whenever they hear a stamp in the collective dance they are drawn back in. When a second ring is close, the first stamp the dancers hear might come from the other ring, in which case they are drawn across. This isn't how it happens in superconductors at all, but it captures the probabilistic nature. OK, so now say one ring randomly chose to coordinate their watches by beginning the second hands at 12 o'clock, while the other ring randomly chose one second later. And imagine the identical dances involve a stamp every ten seconds. Then the pair in the gap is nine times more likely to end up in the ring which stamps first, regardless of which ring they originated in. The reason is this: there are nine seconds where the next stamp will come from the earlier ring, and only one second where the next stamp will come from the later ring.

Josephson junctions remind me of 'magic hills': places where you stop your car and when you release the handbrake, your car either rolls along a flat road or up a slope. They work thanks to an optical illusion in the landscape: the car rolls downhill, but appears to be rolling on the flat or uphill. The Josephson effect, on the other hand, is no illusion. Place two superconductors close by and current flows from one to the other without an applied voltage. By analogy you might imagine identical car parks at either end of a magic road: cars spontaneously roll from one car park to the other without requiring a change in height. If you instead park the cars at the top and bottom of a magic hill, the cars will roll back and forth from one car park to the other, even spontaneously going uphill.

This all seems too good to be true: surely spontaneous electric currents must be forbidden magic. The ability of supercurrents to flow forever without loss seems like it, too, should be impossible. Yet both happen. Seeing how the laws of physics survive such exotic states grants a deeper appreciation of the laws themselves.

Resilient laws

The first law of thermodynamics says that energy cannot be created or destroyed. Now, an electron moving in a circle radiates light; in a quantum description, it emits photons, but either way it loses energy. Yet a supercurrent, a flow of electrically charged Cooper pairs, can travel around a ring for all eternity. Shouldn't the Cooper pairs lose energy? If so, they can't go forever, because they'd eventually run out of steam (if you'll forgive the thermodynamics pun). In fact, even when the supercurrent flows around a ring, the Cooper pairs remain in a state of lowest possible energy. They can't lose energy, by photons or otherwise. So how can this be? Why doesn't the superconductor radiate light?

It's the same reason an orbiting electron doesn't fall into a nucleus, but on a bigger scale. The electron is not really orbiting: if it were, it would radiate energy and fall in. Rather, if the electron is in a state of definite energy, it must be in a quantum superposition of locations. The best you can say is that there's a probability of finding it at any given place. The mysterious thing we saw in Chapter V is that, when you look for the electron, you find it in one place. It's the same for the supercurrent flowing around the ring: Cooper pairs can't be circling or they would radiate energy, and this would contradict the fact that they are in a state of lowest energy to start with. The thing that flows is the probability to find a given pair, not the pair itself.

The second law states that energy converts from useful work to useless heat over time. Now, an electrical current is certainly useful, and it takes work to move a current down a power line. In doing so energy is lost to heat, sound and vibrations. A supercurrent is equally useful, but it is able to travel entirely without loss. Does this not violate the second law? It does not: the resolution is that it takes no work to cause a supercurrent to flow. Work is done to overcome resistance, but supercurrents flow without resistance, and so there is no work to convert to heat.

In both cases I find it helps my understanding to imagine trying to *really* violate the laws. Imagine some interaction causes the Cooper pairs in the supercurrent to clump together. The current flows around the ring as before, but it is now conveyed by a big blob of charge. Well, in that case the blob really would radiate and lose energy in accordance with the first law. Radiation is a form of heat, so this process also enacts the second law. What this tells us is that having the charge clumped together cannot be the state of lowest energy: the charge must be evenly spread out in a supercurrent.

So the laws of thermodynamics survive in harmony with supercurrents. There is no fundamental law of physics stopping them from existing at high temperatures; yet so far supercurrents are confined to the world of the ultra cold. This returns us to the motivating question of this chapter: how do we put supercurrents to practical use in our middle realm?

The philosophers' stone

The dancers in the fairy ring feel an attraction to one another because of the vibrations they create, the attraction through which they bind into pairs. In the BCS theory of superconductivity electrons feel an attraction to one another, originating in the vibrations through the crystal lattice: phonons. There is experimental evidence for this in the form of the 'isotope effect': heavier isotopes of an element, containing more neutrons, behave the same way chemically but have a greater mass; mercury, as Onnes discovered in 1911, is a pure element that superconducts, and heavier isotopes of mercury become superconductors at lower temperatures. This fits with the predictions of BCS theory if it is assumed that phonons mediate the attraction between electrons, because heavier atoms would vibrate more slowly. This is also the reason there is a maximum possible temperature at which superconductors can exist: this temperature, which at atmospheric pressure is about 40 K, is set by the highest-frequency vibrations that are possible.

It was therefore a complete surprise when, in 1986, yttrium barium copper oxide was found to superconduct at over 100 K. The theory said this should be impossible. In fact, the theory said that YBCO (as it is called) shouldn't be a superconductor at all: until then, all known superconductors had been metals under ambient conditions, while YBCO is a ceramic. Imagine you'd asked someone in 1985 to select from their mantelpiece the object most likely to superconduct above 40 K. Everyone would instinctively have reached for the metal carriage clock; but everyone would have been wrong: they would have done better to have reached for the porcelain carp.

Since that groundbreaking discovery the temperature at which superconductivity can be observed has gradually crept up. A variety of totally unexpected materials has now been found to superconduct.

The record is about 150 K. That is significant progress, but has not yet reached the coldest natural temperatures on Earth.

At present, high-temperature superconductors remain a mystery. The search for them is experiment-led: there is an industry of checking every material for its superconducting properties, in the hope of finding the philosophers' stone – a *room temperature* superconductor.

Some progress has been made. In 2020 it was reported that carbonaceous sulphur hydride had been found to superconduct at 15°C – a slightly chilly room temperature. Unfortunately, however, this was only found to occur at a rather impractical 2.6 million times atmospheric pressure. The result has since been called into question, with other groups failing to reproduce it. At the time of writing the matter remained highly controversial.

What we require is a *theory* of high-temperature superconductivity to guide the experimental search. Right now there are many competing theories, but there is nothing like a consensus. Part of the problem is that many bizarre phenomena accompany high-temperature superconductivity, and it is not yet known which of these help it and which hinder it.

While a theory of high-temperature superconductors is currently lacking, they are nevertheless already put to regular use. YBCO superconducts well above 77 K, the boiling point of liquid nitrogen. Nitrogen is readily available and can be condensed to a liquid straightforwardly, which means that any superconductor that exists above 77 K is already relatively practical. YBCO is used for the superconducting magnets in research institutions including the Large Hadron Collider. High-temperature superconductors are already used for commercial power supplies in certain parts of the United States and Germany: plans exist to connect three of the biggest US power grids using liquid-nitrogen-cooled superconductors. By balancing the power load across the nation, this promises to enable the future adoption of renewable energy.

A great deal of progress has been made, but for now we have only fragments of the spell. We have glimpsed the philosophers' stone, but we have not yet laid hands on it, and for now the theoretical explanation of high-temperature superconductors lies in our future. Forty-six years separated the experimental discovery of conventional superconductors from their theoretical explanation; perhaps the time is right for another such breakthrough.

A rigid route across the dunes

Alchemists once sought the philosophers' stone which would transmute base metal to gold and grant its bearer immortality and freedom from the inevitability of loss. More valuable and much rarer than gold is any high-temperature superconductor, which grants freedom from loss in electrical currents. The stone sought by modern philosophers is a room-temperature superconductor. With that, perhaps, we can usher in a world in which technology facilitates balance rather than growth.

It would be disingenuous to claim that we can solve the world's energy problems with technology. Technological solutions are only useful if we change our personal habits and attitudes, and see coordinated governmental action. Fortunately these advances need not be mutually exclusive. As we reach the conclusion of this book we return to the central theme: re-enchantment with the familiar. My guess is that the concerns of the far future will not be about making new things, or bigger or more efficient things, so much as appreciating what is already there.

On this note, superconductors and superfluids have one last lesson to teach us about the familiar world around us. They manifest quantum phenomena in our middle realm, and their behaviour is totally counterintuitive. That's magic, and, specifically, it's the practical hands-on magic of the wizard. But here's the thing: *superconductors are no more quantum than crystals are.* They're just less familiar. All they're doing is exactly the same thing that all matter does: developing rigidity through spontaneous symmetry breaking. Try to stop the flow of a supercurrent, and the Cooper pairs resist collectively. Push a crystal, it moves as a whole. Superfluids escaping Houdini-style from inescapable bottles is magic — but a crystal sitting on a table is *exactly the same type of magic!* This doesn't mean that superfluids are less magical. Rather, it means that all of matter is just as magical as superfluids — it's just that some matter is familiar. If you lived in a much colder world, superconductivity would be as mundane to you as solidity.

Any theory of high-temperature superconductivity may itself lie in the far future. Due to its potential to change our world for the better, obtaining it is a central aim of condensed matter physics. When we understand high-temperature superconductors as well as we understand BCS theory, we will in all likelihood understand the world entirely differently. Such breakthroughs do not come from nowhere, and are not made in isolation: just as states of matter gain their stability from the cooperation of their many constituent particles, the advancement of human understanding is enabled by the cooperation of many individuals. There is a place for you in the search, if you'll join.

IX

The Boundless Vista

In the northernmost of the western islands lives a woman whom the islanders believe to be the greatest living knot master. The other knot makers, on the other hand, know her to be the greatest knot master *who will ever live*. While endowed from a young age with those natural skills evident to the watchers, knot masters must nevertheless learn their art during their lifetime. As a result, most have a specialisation within their field, with the nexus adepts being one example. The master of the northern island belongs to the ancient matrilineal line of byssus weavers, knot makers who dive deep into the ocean in search of certain special oysters. The byssus weavers seed the oysters with grains of sand. Extremely rarely, a giant pearl will result. However, their hope is for a result far rarer still: they seek to seed the oysters such that, over a period of months and years, they grow a few strands of byssus, or sea-silk, a delicate strand-like substance with the lustre of gold. The byssus weavers dive down to harvest the few strands of sea-silk. Back on land, they weave them into beautiful tapestries, to mark significant events or to honour visiting guests of high regard. The byssus is so delicate that uncountable numbers of strands would be needed to achieve the thickness of spider silk. A skilled knot maker can weave byssus tapestries with such detail that they are visually indistinguishable from the scene they depict.

When she is ninety-three years of age, the master of the northern island creates something never created before: not a byssus tapestry, but a byssus nexus. With this act she creates the world. Just as a byssus tapestry can depict a scene with all the perfection and imperfection of reality, the byssus nexus contains within it an entire universe, brought to life as she speaks the knots and makes the web's connections. As she reconnects the nexus, change occurs. Those inhabitants of the byssus nexus who think only in linear sentences perceive such changes as constituting a linearly flowing time. She rests; upon resumption of her reading and knotting, the inhabitants perceive no discontinuity in the flow of their existence, deriving as it does simply from the changing connections of the great byssus web. The entire past and future of the byssus universe is encoded in the web from its inception. This sequence of sounds describes a mountain; this sequence the death of a sparrow; this sequence the after-image of lightning; this sequence the master herself.

With that Veryan closed the book, her mind returning to the bustling library — and thoughts of escape.

Veryan sought to return a once-ubiquitous magic to the world, a classic theme in fiction. Examples appear in many of the works referenced in this book: *The Lord of the Rings* had a more magical age in the past; Hope Mirrlees's *Lud-in-the-Mist*, Lord Dunsany's *The King of Elfland's Daughter*, and Susanna Clarke's *Jonathan Strange and Mr Norrell*

have the return of magic as their central themes; and the idea appears throughout television shows such as *Avatar: The Legend of Korra* and *Arcane*, and in classic films such as Studio Ghibli's *Princess Mononoke*.

The reason for this prevalence seems clear: if you want to believe that the world is magical, you have to explain why magic is no longer apparent. Ancient accounts of the world often contain rather magical occurrences, and so one explanation is that magic used to exist, but disappeared before the modern age. Often the time at which magic disappeared can be located precisely. Mirrlees, Dunsany and Tolkien were all writing in the interwar period when mechanisation was taking over the old ways of life. Tolkien stated that the threat of Sauron's evil mechanising forces on the Shire, the pastoral utopia of the hobbits, represented the expansion of industrialised Birmingham into the surrounding countryside which was his childhood home. This industrialisation was facilitated in part by technological advances during the First World War, whose brutality he had experienced first hand as a soldier. The wars led to massive technological advances, via increased funding and a sense of urgency. Condensed matter physics developed in precisely the same period, and for the same reasons. The understanding and worldview it brought with it might be thought of as the death of magic.

But I would argue that the magic never really left: it just changed form. It's easy to see a certain kind of magic in the natural world: trees, bubbling brooks and moss-covered rocks in ancient caverns are clear examples. That is stage one of understanding magic: enjoying the show. As we came to understand the world around us, it was tempting to think that the magic departed. But to hold such a view is to be stuck in stage two: understanding some of the tricks, and dismissing the show as trivial. But there is magic in our modern world too. The world never really lost its magic, and with our newfound understanding we can reach stage three, appreciating the show with the insight of the professional magician.

There are well-established health benefits to spending time in nature, such as an alleviation of stress and anxiety.[26] But what is it about nature which grants it this power, and what forms must nature take to work this magic? I was recently walking across the remote wilds of Dartmoor, a barren windswept moorland of peat bogs scattered with giant granite monoliths and ancient oak forests, with an old friend and mentor, Damien Hackney. He was explaining an idea he was developing as part of his PhD. The idea consisted of ways to build the connection and sense of peace which is so easily achieved in nature, for people who have only experienced city living: learning to see the magic not in trees and mountain streams, but in concrete and metal. Filtered through the lens of my own understanding, it looked like a quest to see the magic in condensed matter – whether an unhewn lump of ore or the purified metal of a city bench.

The onward journey

Veryan was walled in by the maze of bookcases, and the iron trellis above and below. Her time was short; her adversaries closing in. While they controlled the library they did not create it, and had no love for it. They valued the knowledge it contained only for the power it granted them.

Veryan's reading had begun to awaken in her memories long-since put to sleep. She recalled mornings in the sun spent reading the adventures of Mister Calabash in his youth, learning from the wise Lady Long-Ears, before he became the master of legend. In those moments of dreamlike indifference to the world, she more than once found herself picking absent-mindedly at the threads of reality around the corners of those familiar pages, too young to know it was impossible. Kindling these memories, her experience passed once again out of space and time. She understood anew that space and time are merely convenient fiction and felt once again the weft from which they emerged.

Spotting a loose thread, she teased an opening in the hessian of existence, creating a tunnel to a point outside the library ten minutes later. Veryan had gained the secret knowledge she sought. This part of her journey was coming to an end. But the purpose of knowledge is to be shared and put to use, and it was to this new adventure which she now set out. It was time to escape the library and return to the world.

Some say there is a deep connection between the worldviews we impose on our scientific theories and the prevailing views of society. If that is the case, the rise of condensed matter physics suggests a positive outlook. One subject on which it has a unique perspective is the question of reductionism and emergence. Reductionism seeks to boil down complex phenomena to the simplest possible descriptions, like Sherlock Holmes solving a case by understanding which details to disregard. Emergence is the viewpoint that when many simple things combine the result may be more than the sum of these parts.

The two approaches are complementary, and both will always play important roles in science. What is at odds with emergence is any assertion that the relevant details must always be the *smallest* constituents. An example would be any programme to reduce reality to a description solely in terms of elementary particles and their interactions, to the exclusion of collective phenomena.

An ancient adage has it that pursuit of one extreme returns you inexorably to the other. Understanding the nature of elementary particles required the development of quantum field theory which told us that particles can never exist in isolation: they are surrounded by a sea of potentiality, particle–antiparticle pairs conjured by a universal sleight of hand. Breaking things down was only ever the first stage of building them back up with understood pieces.

The future holds an emerging role for emergence. It is the focus of condensed matter physics, which works with the quantum building blocks of the microscopic world to understand the complex systems that emerge when many of these blocks combine: the middle realm of our everyday existence. The interesting cases are those where the description at the larger scale is fundamentally different from that at the smaller.

It is striking that there are so few books on condensed matter even though it is the largest discipline in physics. Like many condensed matter physicists, I grew up inspired by tales of string theory, black holes and multiverses. It was a pleasant surprise to discover at university a vast and fascinating subject which had been kept hidden all those years. But why is it so underrepresented in the popular imagination? When I've asked physicists and science journalists over the years, their answers have tended to fall along two lines. First, that superstrings and black holes are magical, while matter, being familiar, is not; second, that superstrings and black holes benefit from their lack of immediate practical use, which allows them to be sold on their own merits – on the excitement researchers see in them. Condensed matter physics has been a victim of its own practical applications. Hopefully you will now agree that familiarity and practicality are not at odds with magic after all. Quite the opposite: they are precisely the magic worked by wizards.

The lack of popular exposition of condensed matter physics has given me the privileged position of telling many readers about it for the first time. I have given snapshots of a few scientists working in the field, which hopefully have given some idea of the breadth of backgrounds they represent. It is unfortunate that for much of history science has been the preserve of a small elite. But this time has passed, and we are left guessing at how much further along we would be on our collective journey if science had been better represented by a broader diversity of backgrounds through the ages. If there is one message I have hoped to convey it is that condensed matter physics is open to everyone: anyone can be a wizard, and if it interests you I encourage you to join the community. If you don't feel you fit the stereotype your perspective is all the more necessary to progress the subject into the future.

I was once told that the secret to all stage magic is that the magician must put in more effort than anyone in the audience would believe. This is also the secret to physics: it can be done, and by anyone, but it takes a lot of practice. When you've got the hang of it, people will see you as a wizard, because they can't imagine the work that went into developing your understanding. Creating this sense of wonder has its drawbacks: people are often put off trying to become physicists, because they feel like mathematical skill is something you are either born with or not. I can personally attest that this is not the case: it is learnt with practice.

The idea of science done in isolation is continually pushed on us: the Nobel Prize can be shared by at most three people, despite the hugely collaborative nature of modern science. Discoveries, in this model, are attributed to individuals without thought to the wider environment of human interaction. Such beliefs can only be held if considering oneself as separate to the world being manipulated. To do so is to forget the laws of wizardry, and it is the same philosophy that has encouraged us to see the environment as a consumable re-source separate to ourselves. But there is reason to be hopeful that the shifting trend towards emergence gives us reason to be optimistic of positive change. We are not separate from our environment, but a part of it. This fits with the scientific trend towards emergence, the whole being more than the sum of the parts; and it is undoubtedly the viewpoint of the wizard.

Appendix

This appendix supplies a proof of a statement in Chapter V. Recall the setup. You and two friends each face a separate night-washer in a separate tower (let's number the towers 1, 2 and 3). At midnight, each night-washer will receive a marble delivered by a raven; you and your friends will then have to guess which hand the night-washer has the marble hidden in, and they will each reveal whether you are correct. The washers have declared the following will always be obeyed:

i. Whenever one of you picks the left hand, an odd number of you will be correct.
ii. Whenever three of you pick the left hand, an even number of you will be correct.

This appendix will prove that there can be no secret rule being followed by the night-washers which is compatible with both i and ii; even if they're cheating, they couldn't possibly obey i and ii unless they are somehow able to know what is happening far away, instantaneously.

Denote {A,B} the outcome for the {Left (L),Right (R)} hand. Each is either 'y' for 'yes, there's a marble' or 'n' for 'no, there's not'. Denote with '.' any outcome that could be either result.

In each case we will write a set of three brackets {L,R} {L,R} {L,R}, one bracket for the outcome in towers 1, 2 and 3 respectively. Looking at condition i, we need to consider the three cases where one person of your trio chooses left: either LRR (tower 1 choosing Left, towers 2 and 3 choosing Right), RLR, or RRL; and looking at condition ii we need to consider the case where all three of you choose left: LLL.

Take the first of these cases, LRR. The possible rules compatible with i are:

{*y,.*} {*.,n*} {*.,n*} *(tower 1 correct under LRR)*
{*n,.*} {*.,y*} {*.,n*} *(tower 2 correct under LRR)*
{*n,.*} {*.,n*} {*.,y*} *(tower 3 correct under LRR)*
{*y,.*} {*.,y*} {*.,y*} *(towers 1, 2, 3 correct under LRR).*

In each row, the person in tower 1 has chosen the left hand, and the other two have chosen the right, which is why the left entry in the first bracket is filled, and the right entry in each of the other brackets is filled. Condition i says that an odd number must be correct, so there must be an odd number of 'y' entries along each row. The four rows exhaust the possible ways in which that is true; '.' appears for example in the right-hand rule for tower 1, as we have so far only considered the case where tower 1 chose left. This way, if the three of you choose to inspect hands LRR, an odd number of you must be correct, and condition i is obeyed.

But you could also choose RLR. Whatever the secret rule is, it must also be compatible with this choice, as the rule must cover all eventualities. This constrains the possible secret rules further. Consider the first row above. There are two ways to fill in some of the '.' entries so that this secret rule is also compatible with RLR. The two options are:

{*y,y*} {*n,n*} {*.,n*} *(tower 1 correct under RLR) or*
{*y,n*} {*y,n*} {*.,n*} *(tower 2 correct under RLR).*

Looking at the second of the four original rules, there are again two possibilities:

{*n,y*} {*n,y*} {*.,n*} *(tower 1 correct under RLR) or*
{*n,n*} {*y,y*} {*.,n*} *(tower 2 correct under RLR).*

Looking at the third of the four original rules, there are again two possibilities:

> {*n,n*} {*n,n*} {.,*y*} *(tower 3 correct under RLR) or*
> {*n,y*} {*y,n*} {.,*y*} *(towers 1, 2, 3 correct under RLR).*

And looking at the fourth of the four original rules, there are once again two possibilities:

> {*y,n*} {*n,y*} {.,*y*} *(tower 3 correct under RLR) or*
> {*y,y*} {*y,y*} {.,*y*} *(towers 1, 2, 3 correct under RLR).*

Finally, you could also choose RRL. Of the eight options just listed, the last place is now fixed in each case:

> {*y,y*} {*n,n*} {*n,n*} *(tower 1 correct under RRL)*
> {*y,n*} {*y,n*} {*y,n*} *(tower 3 correct under RRL)*
> {*n,y*} {*n,y*} {*y,n*} *(towers 1, 2, 3 correct under RRL)*
> {*n,n*} {*y,y*} {*n,n*} *(tower 2 correct under RRL)*
> {*n,n*} {*n,n*} {*y,y*} *(tower 3 correct under RRL)*
> {*n,y*} {*y,n*} {*n,y*} *(tower 1 correct under RRL)*
> {*y,n*} {*n,y*} {*n,y*} *(tower 2 correct under RRL) or*
> {*y,y*} {*y,y*} {*y,y*} *(towers 1, 2, 3 correct under RRL).*

The final rule in the list is that considered in the main text, 'everyone always correct'. These eight rules are the only possible rules compatible with condition i, regardless of how they are encoded.

To check compatibility with ii, add up the number of 'y's in the left entries of each pair within a row. The number must be even for condition ii to be obeyed. But the number is always odd. Therefore there can be no secret rule compatible with i and ii, which is what we set out to show.

Acknowledgements

There are many people I wish to thank, without whose help this book would either not exist or would exist in an unreadable form.

First, I wish to thank my agent Antony Topping for guiding me through the entire publishing process, and having an infectiously positive outlook at all times. I also wish to thank Marcus du Sautoy for putting us in touch.

The readability of the book is thanks to my editors Helen Conford, Ed Lake and Nick Humphrey at Profile, and Eamon Dolan at Simon & Schuster. Nick Allen at Forewords provided not only a thorough copy-edit, but also many helpful scientific comments.

I wish to thank Fritjof Capra both for inspiring me with his writing, then later encouraging me with my own.

I wish to thank Femi Fadugba and Eugenia Cheng for a great deal of help, support and advice on the writing and publishing process.

I wish to thank Ruth Gordon for allowing me to use her painting of Maxwell's demon, and for creating the painting in the first place. I wish to thank my stationer and good friend Azeem Zakria of Scriptum in Oxford for crafting me a 'magic journal' in which I could note down observations of everyday magic. I wish to thank Helena Laughton and Sixuan Chen for translating the passage in the *Guiguzi*; given that it is the first-known reference to magnets, a translation turned out to be surprisingly hard to come by.

I wish to thank all of my friends who took the time to read various drafts of the book and give helpful suggestions and comments. In particular, Dina Genkina, John Hannay, Kun Lee, Gwendoline Lindsay-Earley, Ellen Masters, Emma Powell, Leonid Tarasov and Jack Winter. The detailed comments of Holger Haas, Sebastián

ACKNOWLEDGEMENTS

Montes Valencia and Jasper van Wezel led me to totally overhaul the first draft, and the book is incomparably improved as a result; Montes and Jasper provided additional sets of detailed comments on later drafts, and it was only after incorporating these that the book began to resemble something readable.

Finally, I wish to thank my fiancée Beatrice Dominique Scarpa both for her profound help and for being a constant source of inspiration and magic, ever since that day we chanced upon one another while walking through the desert.

Notes

I will sometimes provide references to scientific articles to back up claims I make about recent research. Many of these have been written for specialist audiences, meaning they are not intended to be readable without a PhD in a relevant field. In many cases, though, popular summaries will be available elsewhere by searching for the title of the paper. I will provide digital object identifiers (DOIs) which uniquely identify the articles.

Some references require paid subscriptions to access them. However, almost all modern physics publications are also uploaded in 'pre-print' form to arxiv.org, where they can be accessed freely. The only differences between the pre-print and journal form usually lie in the formatting, not in the content. Note that papers on arxiv.org need not be peer reviewed; if an article on arxiv.org has passed peer review and been published in a journal, the journal reference will usually be given.

CHAPTER I: THE PHYSICS OF DIRT

1 C. A. Moss-Racusin et al., 'Science faculty's subtle gender biases favor male students', *Proceedings of the National Academy of Sciences of the USA*, vol. 109(41), pp. 16474–9 (2012); doi. org/10.1073/pnas.1211286109

CHAPTER II: THE FOUR ELEMENTS

2 N. J. Mlot et al., 'Fire ants self-assemble into waterproof rafts to survive floods', *Proceedings of the National Academy of Sciences of the USA*, vol. 108(19), pp. 7669–73 (2011); doi.org/10.1073/ pnas.1016658108

3   J. L. Silverberg et al., 'Collective motion of humans in mosh and circle pits at heavy metal concerts', *Physical Review Letters*, vol. 110, 228701 (2013); doi.org/10.1103/PhysRevLett.110.228701

CHAPTER III : THE MAGIC OF CRYSTALS

4   See, for example: N. Mirkhani et al., 'Living, self-replicating ferrofluids for fluidic transport', *Advanced Functional Materials*, vol. 30(40), 2003912 (2020); doi.org/10.1002/adfm.202003912

CHAPTER IV: REFLECTIONS ON THE MOTIVE POWER OF FIRE

5   This being a famous case study, the numbers have been rechecked repeatedly; the values I quote are from a methodical 2014 review of Galton's original lists, which contained small errors. The paper can be found here: K. F. Wallis, 'Revisiting Francis Galton's forecasting competition', *Statistical Science*, vol. 29(3), pp. 420–4 (2014); doi.org/10.1214/14-STS468

6   S. Toyabe et. al., 'Experimental demonstration of information-to-energy conversion and validation of the generalized Jarzynski equality', *Nature Physics*, vol. 6, pp. 988–92 (2010); doi.org/10.1038/nphys1821

7   For more on quantum thermodynamics you can check out Nicole Yunger Halpern's book *Quantum Steampunk* (Baltimore, MD, Johns Hopkins University Press, 2022). It, too, interweaves a fantasy narrative with factual science writing.

CHAPTER V: BEYOND THE FIELDS WE KNOW

8   A. M. Ionescu and H. Riel, 'Tunnel field-effect transistors as energy-efficient electronic switches', *Nature*, vol. 479, pp. 329–37 (2011); doi.org/10.1038/nature10679

9   For a discussion of whether this is the case, see: P. Ball, 'Is photosynthesis quantum-ish?', *Physics World*, vol. 31(4), p. 44 (2018); doi.org/10.1088/2058-7058/31/4/39

CHAPTER VI: SPELLS OF DIVISION

10 F.-Q. Xie et al., 'Gate-controlled atomic quantum switch', *Physical Review Letters*, vol. 93, 128303 (2004); doi.org/10.1103/physrevlett.93.128303

11 C. Castelnovo et al., 'Magnetic monopoles in spin ices', *Nature*, vol. 451, pp. 42–5 (2008); doi:10.1038/nature06433

12 F. K. K. Kirschner, F. Flicker, A. Yacoby, N. Y. Yao and S. J. Blundell, 'Proposal for the detection of magnetic monopoles in spin ice via nanoscale magnetometry', *Physical Review B*, vol. 97, 140402 (2018); doi.org/10.1103/PhysRevB.97.140402

13 P. A. M. Dirac, 'Quantised singularities in the electromagnetic field', *Proceedings of the Royal Society of London*, Series A, vol. 133, p. 60 (1931); doi.org/10.1098/rspa.1931.0130

14 R. Dusad, F. K. K. Kirschner, J. C. Hoke, B. R. Roberts, A. Eyal, F. Flicker, G. M. Luke, S. J. Blundell and J. C. S. Davis, 'Magnetic monopole noise', *Nature*, vol. 571, pp. 234–9 (2019); doi.org/10.1038/s41586-019-1358-1

15 C. Kim et al., 'Observation of spin–charge separation in one-dimensional $SrCuO_2$', *Physical Review Letters*, vol. 77, 4054 (1996); doi.org/10.1103/PhysRevLett.77.4054

16 R. McDermott et al., 'Microtesla MRI with a superconducting quantum interference device', *Proceedings of the National Academy of Sciences of the USA*, vol. 101(21), pp. 7857–61 (2004); doi:10.1073/pnas.0402382101

17 J. Hong et al., 'Experimental test of Landauer's principle in single-bit operations on nanomagnetic memory bits', *Science Advances*, vol. 2(3), e1501492 (2016); doi.org/10.1126/sciadv.1501492

CHAPTER VII: SPELLS OF PROTECTION

18 Actually there are two articles. The first is N. D. Mermin, 'Is the Moon there when nobody looks? Reality and the quantum theory', *Physics Today, vol. 38(4), 38 (1985); doi.org/10.1063/1.880968*. This

section is based also on a follow-up: N. D. Mermin, 'Quantum mysteries revisited', *American Journal of Physics*, vol. 58, 731 (1990); doi.org/10.1119/1.16503

19 A. Einstein et al., 'Can quantum-mechanical description of physical reality be considered complete?', *Physical Review*, vol. 47(10), pp. 777–80 (1935); doi.org/10.1103/PhysRev.47.777

20 J.-W. Pan et al., 'Experimental test of quantum nonlocality in three-photon Greenberger–Horne–Zeilinger entanglement', *Nature*, 403(6769), pp. 515–19 (2000); doi.org/10.1038/35000514

21 J. Yin et al., 'Satellite-based entanglement distribution over 1200 kilometers', *Science*, vol. 356(6343), pp. 1140–44 (2017); doi. org/10.1126/science.aan3211

22 There were papers by two competing groups. The first was H. Bartolomei et al., 'Fractional statistics in anyon collisions', *Science*, vol. 368 (6487), pp. 173–7 (2020); doi/10.1126/science.aaz5601. The second was J. Nakamura et al., 'Direct observation of anyonic braiding statistics', *Nature Physics*, vol. 16(9), pp. 931–6 (2020); doi.org/10.1038/s41567-020-1019-1

CHAPTER VIII: IN SEARCH OF THE PHILOSOPHERS' STONE

23 United States Energy Information Administration (EIA), *Monthly Energy Review*, April 2022. The report indicates that transmission and distribution losses for 2021 were 226 billion kWh (about 5.5% of the total electrical energy generated). In May 2022 electrical energy cost 13.83 cents per kWh. The New York City Department of Transport maintains 315,000 streetlights as of May 2022, with a typical LED streetlight consuming 80 W (road lights use around 150 W, while residential streetlights use 35 W).

24 C. E. Tarnita et al., 'A theoretical foundation for multi-scale regular vegetation patterns', *Nature*, vol. 541, pp. 398–401 (2017); doi. org/10.1038/nature20801

25 Y. Ronen et al., 'Charge of a quasiparticle in a superconductor',

Proceedings of the National Academy of Sciences of the USA, vol. 113(7), pp. 1743–8 (2016); doi.org/10.1073/pnas.1515173113

CHAPTER IX: THE BOUNDLESS VISTA

26 See, for example: M. P. White et al., 'Spending at least 120 minutes a week in nature is associated with good health and wellbeing', *Scientific Reports*, vol. 9, 7730 (2019); doi.org/10.1038/s41598-019-44097-3. G. N. Bratman et al., 'Nature experience reduces rumination and subgenual prefrontal cortex activation', *Proceedings of the National Academy of Sciences of the USA*, vol. 112(28), pp. 8567–72 (2015); doi.org/10.1073/pnas.1510459112. V. F. Gladwell et al., 'The great outdoors: how a green exercise environment can benefit all', *Extreme Physiology & Medicine*, vol. 2, 3 (2013); doi.org/10.1186/2046-7648-2-3

Index

Note: Italic page references indicate an illustration; the suffix n, a footnote.